Gene structure and expression

GENE STRUCTURE AND EXPRESSION

SECOND EDITION

John D. Hawkins

Published by the Press Syndicate of the University of Cambridge
The Pitt Building, Trumpington Street, Cambridge CB2 1RP
40 West 20th Street, New York, NY 10011–4211, USA
10 Stamford Road, Oakleigh, Victoria 3166, Australia

First published 1985
Second edition 1991
Reprinted 1991, 1993

Printed in Great Britain at the University Press, Cambridge

British Library Cataloguing in publication data

Hawkins, John D.
Gene structure and expression.–2nd edn
1. Organisms. DNA
1. Title
574.873282

Library of Congress cataloguing in publication data available

ISBN 0 521 39230 6 hardback
ISBN 0 521 39855 X paperback

UP

Contents

Introduction to the second edition

Since the first edition of this book appeared there have been many advances in our understanding of the genome, so it is opportune to review some of these in this new edition.

New techniques continue to be invented, and those that have come into general use are described in Chapter 3 on methodology. Similarly, new kinds of vectors are described in Chapter 4. Chapter 7 on eukaryotic gene organisation and expression has been completely re-written as it is in these fields that some of the most dramatic advances have occurred in the past five years. The chapter on hormone genes has been transmuted into a chapter on gene families since much information is now available about their evolution. In Chapter 12 it has been possible to include new material on chloroplast genomes. Elsewhere there have been some re-arrangements and updating of material.

I am most grateful to Dr Adam Wilkins at CUP who asked me to prepare a new edition and to Dr Paul Lasko of the Department of Genetics at Cambridge University. They have both made some very helpful comments and suggestions after reading the first draft. Dr Robin Smith, also of CUP, saw the final version through the press and also made some valuable suggestions for which I am very thankful.

Finally, I should like to thank colleagues at Barts, especially Professor Gavin Vinson and Dr Ian Phillips, for much encouragement and helpful discussion, and, as ever, my wife for her forebearance while I have been working on the book.

Introduction to the first edition

There has been an explosive growth in our detailed knowledge of genetics at the molecular level over the last few years, and it is likely that accretion of new knowledge will occur at an ever increasing rate. It is therefore very difficult even for the specialist to keep abreast of all the latest ideas which rapidly progress from hypothesis to theory to accepted dogma. In the time that it takes to write a comprehensive text book it is inevitable that new ideas will be generated and many problems in the field elucidated so that such a book will certainly be out of date before the writing is finished, let alone published. Even during the writing of this small book, over the course of a little more than a year, much new information has come to light so that were it to be re-written in the next few months, appreciable differences would appear. It does not therefore claim to be a complete guide to the subject under review; nevertheless it attempts to present ideas that are reasonably well established and at the same time to cover a fairly wide field, albeit mostly not in great depth. The selection of topics as examples of our knowledge is somewhat arbitrary and conditioned by the author's own interests and expertise.

I believe that it should be a useful book for medical students who wish to become familiar with recent ideas and techniques in molecular biology to help in understanding further advances when they arrive. It will also be of use to honours and graduate students in genetics, biochemistry and those who would not necessarily regard the topics discussed here as their major interests in these subjects. It assumes a working knowledge of biochemistry that a first or second year university or polytechnic student should have acquired in a fairly elementary course in that subject. This basic material is already excellently covered in such books as Lubert Stryer's *Biochemistry* and Albert Lehninger's *Biochemistry*, as well as a host of others.

Reading Lists for each Chapter are to be found at the back of this book. They are mostly made up of review articles in which references to original work can be found. In several cases parallel reviews covering more or less the same ground have been cited.

I am most grateful to Dr Fay Bendall of Cambridge University Press who encouraged me to write this book in the first place, widening my horizons and giving me a good deal of pleasure in the process: to Dr Audrey O. Smith whose patience and skill in pointing out errors and inconsistencies in the text have helped to clarify it. Needless to say, any errors that do remain are my own responsibility. I am also grateful to many colleagues at St Bartholomew's Hospital Medical College, particularly Dr Clem Lewis for stimulating discussions which I have frequently found helpful in clarifying my ideas. Last, but not least, I am grateful for my wife's forebearance in the face of a neglectful husband who has been preoccupied for many months with the genome rather than with its ramifications as revealed in family life.

June 1984

Abbreviations

In general, standard biochemical abbreviations are used throughout this book, particularly in figures and tables. The following is a list of those that are used without further explanation.

Amino acids:

Alanine Ala, A
Arginine Arg, R
Asparagine Asn, N
Aspartate Asp, D
Cysteine Cys, C
Glutamine Gln, Q
Glutamate Glu, E
Glycine Gly, G
Histidine His, H
Isoleucine Ile, I
Leucine Leu, L
Lysine Lys, K
Methionine Met, M
Phenylalanine Phe, F
Proline Pro, P
Serine Ser, S
Threonine Thr, T
Tryptophan Trp, W
Tyrosine Tyr, Y
Valine Val, V
Termination Ter (at end of protein sequences)

Purine and pyrimidine bases and other symbols used in writing DNA or RNA sequences. It should be clear from the context whether bases or nucleotides are intended:

Adenine A
Guanine G
Cytosine C
Uracil U
Pseudo-uridine Ψ
Thymine T
Pyrimidine Y
Purine R
Any nucleotide N
Any base B
Phosphate radical P
Ribose R
Deoxyribose dR
Methyl group m

1
DNA

1.1 The genetic material

The classic experiments of Avery in 1944 demonstrated that DNA (Deoxyribonucleic acid) is the material that can pass genetic information from one bacterium to another. He showed that strain-specific properties of related bacteria could be transferred by DNA that was free of proteins and other substances. DNA is a polymeric molecule built up from only four similar but distinct monomers – nucleotides which are the 5′-phosphates of deoxyguanosine (dGMP), deoxyadenosine (dAMP), deoxycytidine (dCMP), and thymidine (TMP) (Fig. 1.1). In DNA these are joined by phosphodiester linkages between the 3′- and 5′-positions of successive deoxyribose moieties. The initial letters of the bases in the nucleotides are used as abbreviations when writing out sequences in DNA.

1.2 DNA is a polar molecule

One end of a DNA molecule has a phosphoryl radical on the C-5′ of its terminal nucleotide, while the other end possesses a free -OH on the C-3′ of its nucleotide. Thus a polynucleotide possesses *polarity* in an analagous way to the more familiar polarity of proteins with free -NH_2 and -COOH groups at each end. This means, for example, that the tetranucleotides TCGA and AGCT are different chemical entities with distinct properties, even though they behave in a very similar way in many respects (Fig. 1.2). By convention, sequences of DNA are written with the nucleotide containing the free phosphoryl radical at the left. Sequences to the left of a given nucleotide are said to be on the 5′-side (often called upstream), and those to the right are said to be on the 3′-side (often called downstream). The symbols N, R and Y are used to denote any nucleotide, a purine nucleotide, and a pyrimidine nucleotide respectively.

1.3 DNA generally exists as a double helix

DNA generally exists in double strands because of the propensity of the bases for hydrogen bonding to each other in a highly specific way (Fig. 1.1). A bonds with T, and G with C, though very occasional mismatches or alternative bonding can occur. Thus a double-stranded DNA will always contain equal molar proportions of A and T and of G and C though the content of A (or T) and G (or C) varies widely in DNA from different sources.

Fig. 1.1. The four deoxyribonucleotides that make up DNA, showing how the bases form hydrogen bonds.

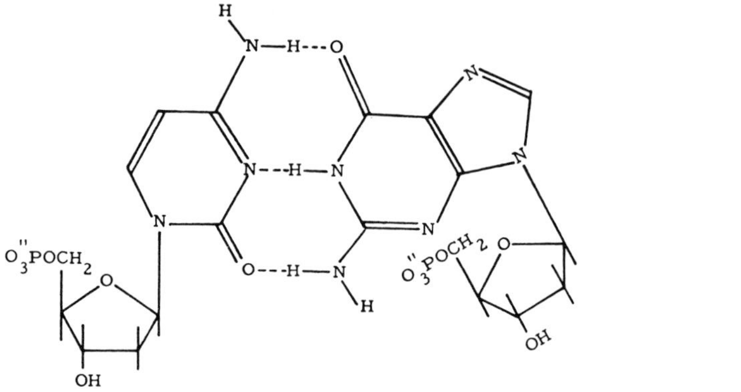

Deoxycytidine-5'-monophosphate
dCMP

Deoxyguanosine-5'-monophosphate
dGMP

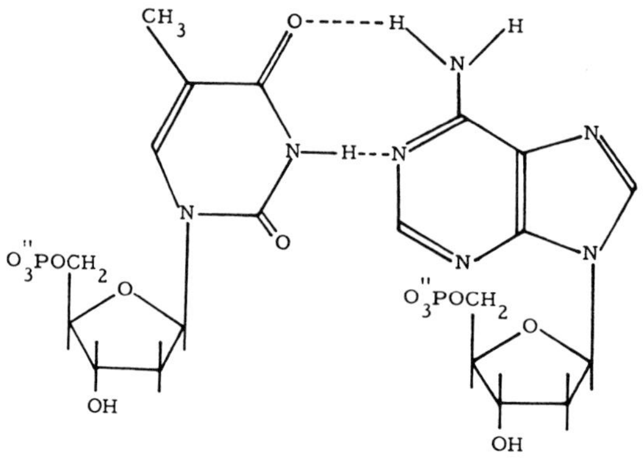

Thymidine-5'-monophosphate
dTMP

Deoxyadenosine-5'-monophosphate
dAMP

The basic unit in a DNA molecule is the pair of nucleotides hydrogen bonded to each other, which is generally known as a base-pair (bp; with kbp used as an abbreviation for 1000 bp).

This double-stranded molecule takes up a helical conformation in which the continuous deoxyribose–phosphate strands twine round the outside of the *helix* with the base pairs (A and T or G and C) in the interior (Fig. 1.3).

The association of A and T with two hydrogen bonds is less stable than that of C and G which has three hydrogen bonds. This has important

Fig. 1.2. Polarity in DNA. The two tetranucleotides TCGA (left) and AGCT (right) are different, even though they have the bases in the same order.

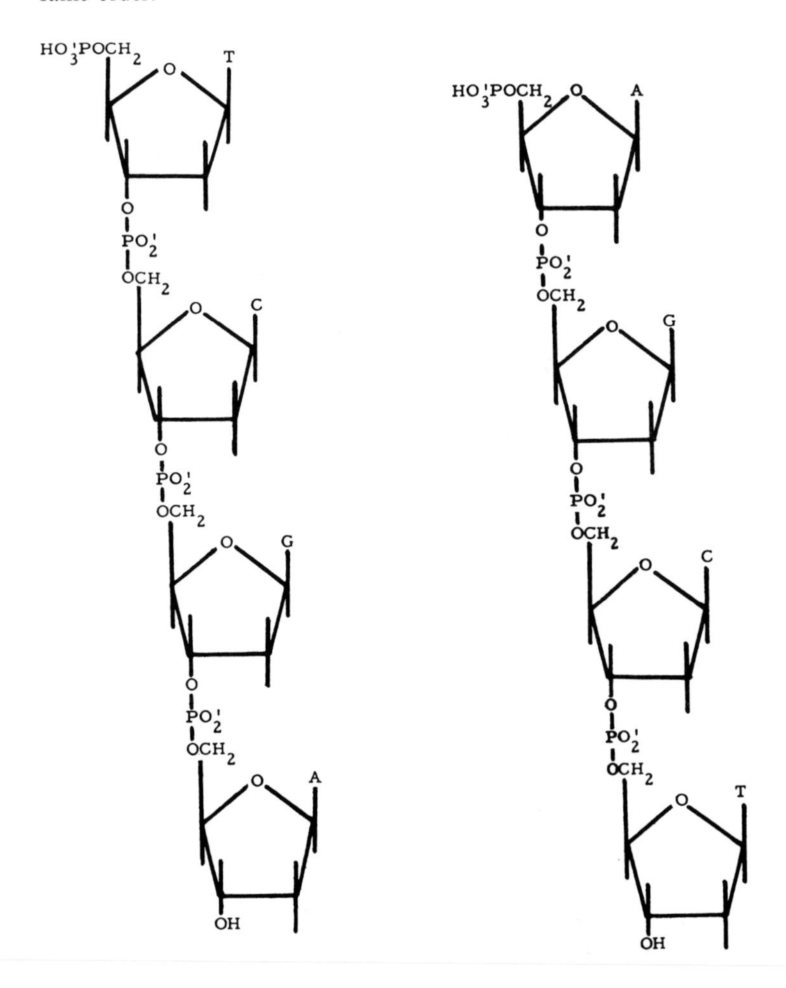

consequences for the stability of different regions of the double helix. In regions that are rich in A and T residues, the helix can be more easily destabilised and unwound than in G–C rich regions.

The polarity of the two DNA strands is *anti-parallel* – that is to say, one runs in the 5′ to 3′ direction while the complementary strand runs the opposite way. The helix can adopt several conformations. The commonest form (B-DNA) has a pitch of just over 10 residues per turn, and is right-handed when viewed end on. Another form, known as Z-DNA, can arise under certain conditions when there are alternating purine and pyrimidine residues in the sequence. This is a left-handed helix and has a pitch of 11.5 residues per turn (Fig. 1.3). This form is believed to have some biological significance, since short stretches of alternating purine and pyrimidine residues occur at numerous sites in many DNAs. They can be detected by the binding of antibodies which specifically recognise this form of DNA. Other non-antibody proteins have been discovered which also react with Z-DNA in a wide range of cells in higher organisms, insects, bacteria and viruses. These viral sequences occur in regions that are known to be involved in the control of the genome, so it is likely that they have important functions in this process.

The geometry of β-DNA is such that there are two grooves with different dimensions running helically along it. The *major groove* (which is

Fig. 1.3. A length of double helical DNA, containing 20 bp, showing both B and Z forms. The lines running round the outside represent the backbone of poly deoxyribose phosphate, while the horizontal lines represent the edges of the base pairs in the interior of the molecule. (Reprinted, with permission, from S. B. Zimmerman, *Ann Rev. Biochem.*, **51** © 1982 by Annual Reviews Inc.)

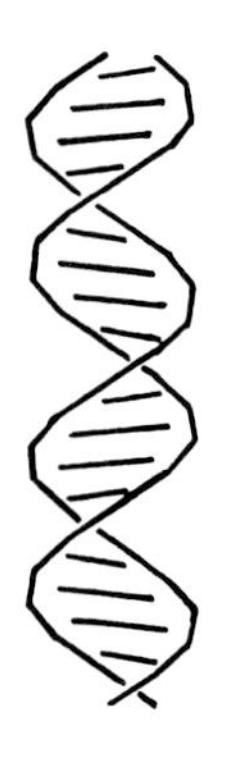

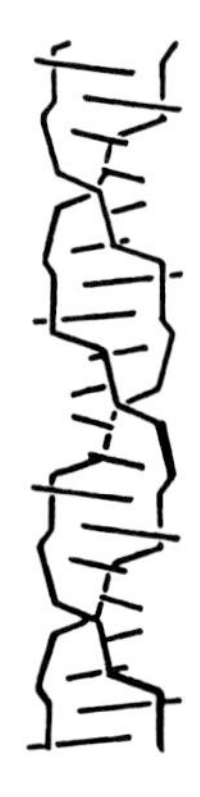

B–DNA

Z–DNA

Table 1.1. *Chromosome numbers and DNA content of cells of a representative set of species*

Species	Number of chromosomes	DNA content kilobase pairs
Bacillus subtilis	1	2×10^3
Escherichia coli	1	3.8×10^3
Saccharomyces cerevisiae	34	14×10^3
Drosophila melanogaster	8	2×10^5
Sea urchin	52	1.6×10^6
Frog	26	45×10^6
Chicken	78	2.1×10^6
Mouse	40	4.7×10^6
Human	46	5.6×10^6
Maize	20	30×10^6

All figures are for diploid cells, except for bacteria.

to the top of Fig. 1.1) is about 12 Å wide and 8·5 Å deep, while the *minor groove* (at the bottom of Fig. 1.1) is only 6 Å wide and 7·5 Å deep. The N and O atoms of the bases and the H atoms of the amino groups lining each groove can serve as hydrogen bond acceptors and donors for making specific contacts with appropriate atoms in DNA binding proteins that control the replication and transcription of DNA (Chapters 5 and 7). Since the overall diameter of the α-helix of a protein is about 12 Å, this can fit snugly into the major groove, provided that suitable complementary atoms are present to make hydrogen bonds between the two molecules.

In addition to hydrogen bonding between the complementary bases, there are individually weak interactions between adjacent rings of the purine and pyrimidine bases that contribute significantly to the stability of the structure. These are known as *stacking interactions*.

1.4 DNA molecules are very long but can be twisted into compact forms

DNA molecules are extremely long and can be visualised by the electron microscope. A double helix of 10^6 base pairs is 0·34 mm long and only 2 nm in diameter. In prokaryotes the DNA is circular so that there are no free 3′- and 5′-ends, and all the chromosomal DNA is in a single molecule. In eukaryotes the DNA in the chromosomes exists as linear molecules, and different species possess different numbers of chromosomes. The amount of DNA in cells of different species varies very widely, generally with an increase in the DNA content as species become more complex (Table 1.1).

With one exception all eukaryotic chromosomes are paired, with one partner coming from each parent. The exception is the sex chromosome. Females carry two X chromosomes, while males carry an X chromosome inherited from their mother and a Y chromosome from their father. This is the case in mammals and many other orders, but other methods of sex determination do occur. The two chromosomes of a pair are said to be *homologous* since they will nearly always be identical in their organisation and frequently in the genes they carry. However, since there are many mutant genes in a population a pair of homologous chromosomes may carry different genes at particular loci. These are known as *alleles*.

A mutant gene encoding a defective product can generally be *complemented* by a 'good' copy of the gene on the homologous chromosome, but if there is a defect on the single copy of the X chromosome that a male carries it cannot be complemented in this way. Thus there are many sex-linked inherited diseases which are carried by females, but expressed only in males. These disorders can actually occur in females but the chances of a female inheriting two defective genes are very low.

Since all somatic cells contain a homologous pair of each of the chromosomes they are known as *diploid*. The gametes – sperm and ova – which only contain one member of each pair of chromosomes are known as *haploid* cells. The contribution of one parent to the genetic make-up of the offspring is known as the *haplotype*.

Individual chromosomes are morphologically distinguishable when they are suitably stained. For purposes of identification they have been given numbers in order, starting with the largest one.

In the cell, both linear and circular molecules are found in much more compact forms. The helix is coiled on itself several times (like the element in an electric light bulb) so that the overall length is greatly reduced at the expense of an increase in diameter. This conformation is stabilised by proteins in eukaryotic cells (Chapter 7.2). This results in packaging DNA into a minimum of space, but when portions of the DNA become functional there is some uncoiling of this structure accompanied by temporary separation of the two helical strands.

In a circular DNA molecule containing 4000 bp (such as might occur as a bacterial plasmid – Chapter 4.1) the double helix is in the B-form. As this has a pitch of 10 residues per turn there should be 400 turns. In practice such DNA is found to have only about 380 turns because the helix is untwisted to a certain extent. This is known as negative *supercoiling*, and is an important feature of the structure of DNA in bacterial cells. It results in a puckered form of the molecule which gives it a more compact structure, and also places considerable torsional strain on it. An analogy can be made by twisting a rubber band held firmly at two diametrically opposite positions. A molecule in which there is no supercoiling is said to be *relaxed*, and there is a dynamic balance between relaxed and supercoiled forms of DNA as a result of the action of two classes of enzymes called *topoisomerases I* and *II* which catalyse the production of one form from the other.

When a DNA molecule is supercoiled it migrates more rapidly on electrophoresis than when it is relaxed. A family of otherwise identical DNA molecules with different degrees of supercoiling can be made visible as a ladder of bands by this technique (Fig. 1.4). Supercoiled molecules appear more compact than relaxed ones when viewed in the electron microscope.

Formation of the supercoiled form in the cell is brought about by an

Fig. 1.4. The effect of supercoiling on the electrophoretic mobility of Simian virus 40 DNA. The DNA was treated with a class II topoisomerase and then electrophoresed. The thick band at the bottom represents fully relaxed DNA, while molecules with increasing degrees of supercoiling appear as bands of increasing mobility. The arrow shows the direction of electrophoretic migration. (Reproduced from W. Keller, *Proc. Natl Acad. Sci., USA* (1975), **72**, 2550.)

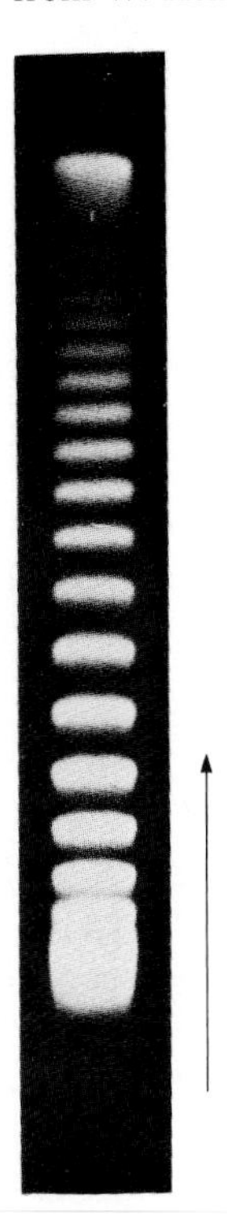

enzyme called DNA *gyrase* (a class II topoisomerase) and, because it is a strained structure, there is an energy requirement for this reaction, which is met by the concomitant hydrolysis of ATP. The reaction proceeds by breaking both strands of the DNA so that another part of the molecule can be passed through the break. The broken strands are then re-sealed (Fig. 1.5). The reverse reaction, in which supercoiled DNA is converted to the relaxed form, is catalysed by a class I topoisomerase and requires no input of energy since a less-strained molecule is being produced. This reaction involves nicking one strand of the DNA, when the 5′-phosphate at the break point becomes bound to a tyrosyl residue on the enzyme. This strand also has a free 3′-OH group and this end is rotated round the other strand which is intact. The nicked strand is then re-sealed (Fig.1.6).

These topoisomerases play important roles in replication and transcription (Chapters 5 and 7). Topoisomerase I is associated with transcriptionally active loci on *Drosophila* chromosomes. Topoisomerase II is an abundant nuclear protein that is localised at the base of DNA loops (Chapter 7.3) where it is believed to be part of the scaffolding on which the DNA is held.

When DNA is to be replicated (see next section) or transcribed into

Fig. 1.5. The action of DNA gyrase in forming a negative supercoil in a circular molecule of DNA. By convention, when the upper strand crosses above the lower strand from left to right the supercoiling is said to be positive. Negative supercoiling is the converse of this.

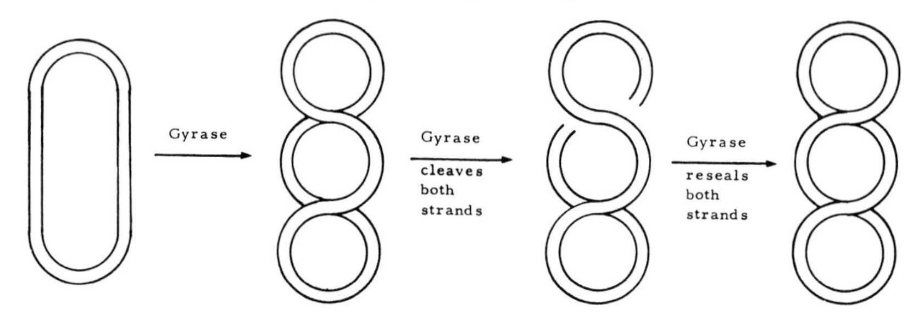

Fig. 1.6. Relaxation of supercoiled DNA by topoisomerase I.

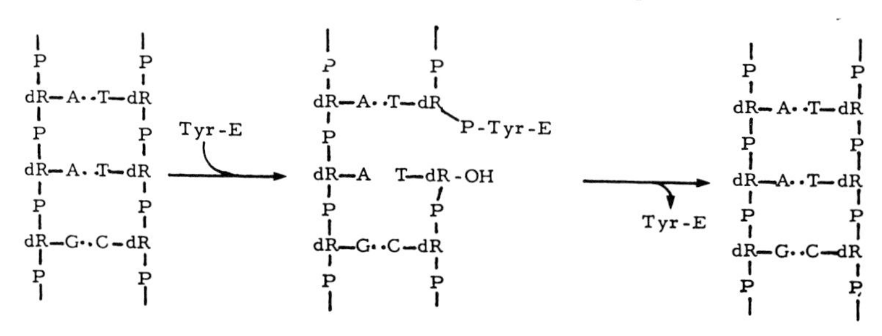

RNA (Chapter 2.2) the double helix must be opened up since each of these processes requires single-stranded DNA as a template. This is performed by enzymes called *helicases* that can produce single-stranded DNA from the duplex form in an ATP dependent reaction. These regions of DNA are stabilised and prevented from re-association by single-stranded binding proteins that bind co-operatively to single-stranded DNA.

1.5 Replication is semi-conservative

When a cell divides, each of the daughter cells contains a full complement of DNA, identical to that of the parent (except during the production of gametes in eukaryotes). Thus the DNA must be precisely *replicated.* This is done by the separation of the two strands, followed by pairing of deoxyribonucleoside triphosphates through specific hydrogen bonding with the bases in each strand. These triphosphates are joined together (ligated) by the enzyme DNA polymerase with the release of inorganic pyrophosphate (Fig. 1.7). Synthesis always proceeds from the 5′-end of the growing chain. In practice the two parental strands do not separate completely, but are opened up at what is known as the *replication fork* which is moved along as the process proceeds.

Fig. 1.7. Replication fork, showing strand separation and incorporation of deoxyribonucleoside triphosphates. Continuous synthesis takes place on the lower strand. Synthesis on the upper strand, which is discontinuous, does not start so near the replication fork.

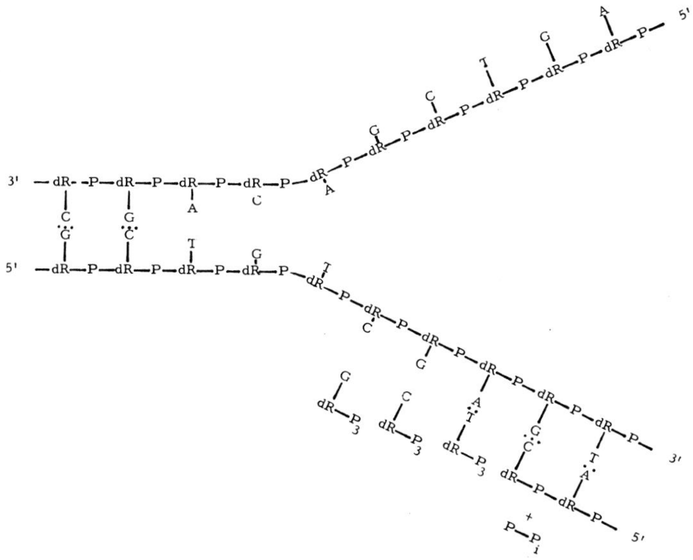

Because of the opposite polarity of the two parental strands, one strand is in the incorrect orientation for continuous synthesis of the new strand. On this strand, comparatively short lengths of new DNA are formed (Okazaki fragments, named after their discoverer) which are then joined by a ligase (Fig. 1.8).

This mode of replication is called *semi-conservative* because one of the DNA strands is conserved. This means that each daughter cell possesses one new strand of DNA and one derived directly from its parent.

This outline of replication may sound very simple, but the actual process is extremely complex and by no means completely understood. In *Escherichia coli*, where it has been most intensively studied, at least 14 different polypeptides, some of which are enzymes, are known to be involved. Further details will be found in Chapter 5.

1.6 The gene or cistron is the functional unit of DNA

Replication ultimately involves the whole DNA molecule, but the functional unit in DNA is much smaller than this – perhaps only a few thousand or even hundred base pairs. These units correspond to the original concept of *genes* postulated by earlier geneticists who associated

Fig. 1.8. Continuous and discontinuous synthesis of two strands of DNA during replication. Note that the scale is much smaller than in Fig. 1.7.

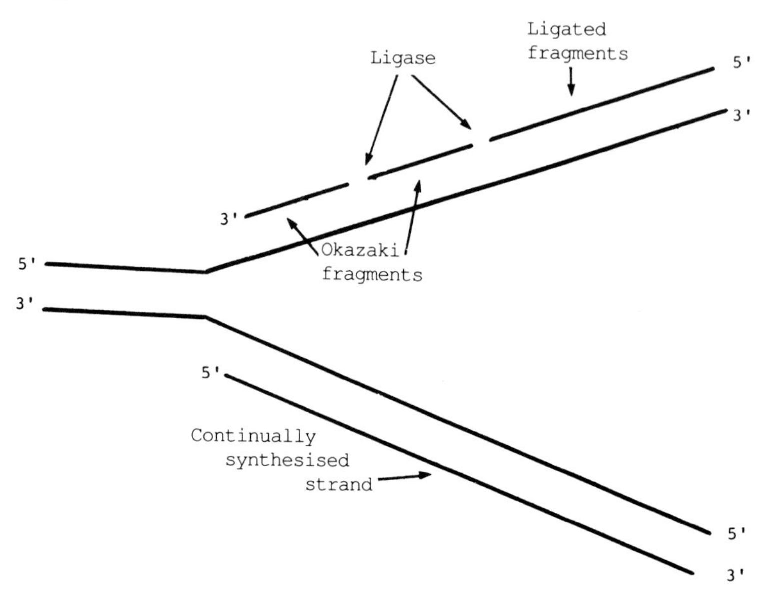

them with various visible (*phenotypic*) characters which could be inherited in predictable ways. A more recent term with a very similar meaning is *cistron*, but this is not widely used, and we still talk of genes. The complete collection of genes in one organism is referred to as the *genome*.

Genes contain information which directs the synthesis of other molecules in a highly specific way. Although only four different nucleotides are involved in building a DNA molecule the number of possible arrangements is extremely large. In a piece of DNA containing 100 nucleotides there are 4^{100} ($1{\cdot}7 \times 10^{60}$) possible sequences. As the genetic material of any organism contains millions of nucleotides there is obviously room to account for all the amazingly wide diversity of living organisms.

Most of the information encoded in the genome is used to specify the sequence of amino acids that are made into proteins to serve a wide variety of essential functions in the cell. As discussed more fully in Chapter 2, the expression of this information is mediated by molecules of RNA, but some RNA is itself used directly for various specific purposes.

However, DNA sequences flanking both ends of those which encode information for making RNAs also have important functions in regulating the activity of genes. In prokaryotes such factors as the availability of metabolites will determine which genes are active at any particular time. In metazoans there is generally specialisation of cells of various types so that some genes in certain tissues are permanently inactive ('switched off'), while others may be activated or not depending on particular local environmental signals.

There are also a number of *pseudogenes* scattered throughout most genomes, with very considerable sequence homology to functional genes from which they have probably been derived by mutations that render them non-functional. They have frequently been excised from the chromosome and re-inserted elsewhere during their genesis, though sometimes they are in their original position.

Finally, there are often long stretches of DNA to which no definite function has so far been assigned, particularly in eukaryotes. In some cases they may be spacer elements between genes encoding specific proteins or RNAs. They may also contain regulatory elements that have not yet been defined, or be involved in chromosome structure. They could possibly provide a pool of DNA from which new genes can arise by mutation.

1.7 Mutations can arise in various ways

The fact that offspring generally resemble their parents very closely means that the constitution of DNA must be fairly stable. However, changes in base sequence can and do occur, though at a rate that is very slow in relation to human life span. If this were not so, evolution could never have occurred. These changes are known as *mutations* and arise as a result of several different mechanisms.

A change in a single nucleotide in a sequence coding for a particular protein or RNA gives rise to a *point mutation.* This may occur when a chemical reaction affects one of the functional groups on a base. Agents altering the structure of the bases in DNA are called mutagens. Some pollutants of our environment may act in this way. A specific example is the deamination of cytosine to uracil resulting from the action of nitrous acid on DNA. (For their structures, see Fig. 2.1.) If cytosine is deaminated to uracil, adenine and not guanine will be incorporated into the new DNA formed during the next round of replication. Mutations resulting in a change of one pyrimidine or one purine to another are known as transitions: changes of one sort of base to the other are known as transversions.

Highly reactive free radicals can be formed by the action of electromagnetic radiation of various wavelengths (e.g. ultraviolet or X-rays). These can cause chemical changes in the bases and break DNA chains. Although enzymes can repair them they do not always function quite correctly. This may lead to either deletion or insertion of one or more nucleotides.

Finally, more drastic changes in the DNA may be brought about by the *translocation* of whole segments from their original site to a new one, either within the same chromosome or on a different one. A special case of this is the phenomenon of *crossing over* which was postulated as a result

Fig. 1.9. Gene conversion and crossing over. ABC and abc are reasonably homologous regions of either one chromosome or two different ones.

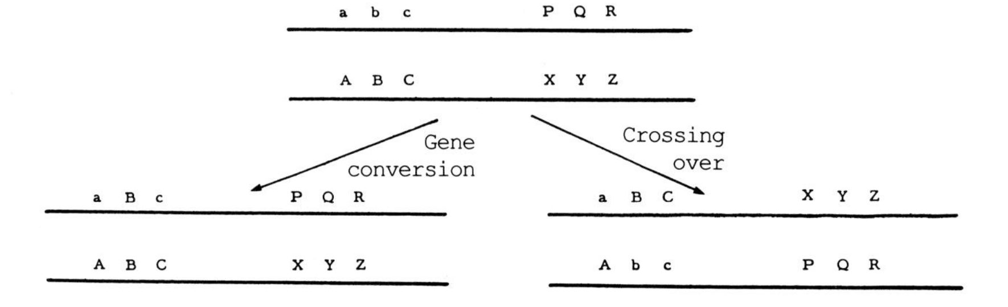

of classical genetic experiments long before the actual mechanism was kown. Crossing over involves reciprocal changes between two homologous chromosomes. The consequence is that two heritable characteristics which were formerly expressed simultaneously will no longer appear together.

Gene conversion is the name given to another important transformation of DNA in which two genes interact so that part of the nucleotide sequence of one is incorporated into the other one. Both genes retain their integrity and location, but a non-reciprocal change in the structure of one of them occurs. Gene conversion probably occurs mostly at meiosis or mitosis, and results from a misalignment of sequences that are normally not paired but which are sufficiently homologous to allow base-pairing to occur (Fig. 1.9). Such an intermediate can give rise either to a cross-over or to transfer of DNA in one direction only (i.e. gene conversion). Gene conversion involves breaks in single strands of DNA, whereas in crossing over both strands must be broken. Gene conversion can occur between genes on different chromosomes (both homologous and non-homologous), or between genes located on the same chromosome. It is probably more common in the latter situation, especially where there are families of reiterated genes of very similar structure, such as the globin genes (Chapter 9); V-region genes of immunoglobulins (Chapter 10); and some genes of the major histocompatibility complex (Chapter 10).

By studying differences in the amino acid sequence of analogous proteins in two different species (e.g. haemoglobins or cytochromes), or, better still, the DNA sequences of their genes, it is possible to make an estimate of the natural mutation rate if other evidence for the time at which the two species diverged is available. The best estimates suggest that on average a mutation occurs about once in 10^6 cell divisions. This sounds a high rate, bearing in mind that there are 10^{14} cells in an adult human, but only those mutations that actually occur in the gametes will be heritable. It is also important to appreciate that many mutations are silent – that is to say they have no observable effect, either because they have occurred in parts of the DNA that do not encode vital information, or even because they have no effect on the encoded information (see Chapter 2.4).

2

Ribonucleic acid

2.1 Expression of the information in DNA is mediated by RNA

Genetic expression always involves the synthesis of another nucleic acid polymer called ribonucleic acid (RNA). RNA is composed of four nucleotide monomers – the monophosphates of adenosine (AMP), guanosine (GMP), cytidine (CMP), and uridine (UMP) (Fig. 2.1). These all contain the sugar ribose instead of deoxyribose found in DNA, and the individual nucleotides are linked together through 3′-, 5′-phosphodiester bridges, just as in DNA. Another major difference from DNA is the presence of uracil rather than thymine as one of the bases. These two compounds differ by the presence of a methyl group in thymine, which is absent from uracil. RNA is generally a single-stranded molecule though its bases can pair by hydrogen bonding to give hairpin-like or stem structures (Fig. 2.2). Guanine occasionally pairs with uracil, though this pairing is less stable than the more usual A–U and G–C pairs. RNA molecules are much smaller than DNA molecules, and only comparatively short stretches of DNA are used to direct the synthesis of individual RNAs.

The stability of stem structures can be calculated in terms of the energy required to open them up. To a first approximation, this is the sum of the energies needed to break the individual hydrogen bonds in the base pairs of the hairpin, but they have to be considered in adjacent pairs because there is considerable dependence on neighbouring bases. An allowance also has to be made for loops at the end of a stem and for bulges in the stem where there may be one or two bases that will not pair with bases on the opposite side of the stem. The free energy for breaking the hairpin structure in Fig. 2.2 is $-27{\cdot}6$ kcal/mole, so it is a stable structure, as might be expected from the high G and C content.

2.2 Transcription is a major stage of gene expression

RNA is synthesised on a template of DNA by a process called *transcription*, in which the strands of the DNA must first separate in the region where it is going to start. As transcription proceeds the DNA strands separate in front of the growing RNA chain. Transcription is formally rather similar to replication, except that only one of the DNA strands is used as a template. Analogously, the immediate precursors of RNA are the nucleoside triphosphates (ATP, GTP, CTP, UTP) whose bases are hydrogen bonded to the complementary bases on the DNA. The enzyme involved is DNA-dependent RNA polymerase. It catalyses the formation of phosphodiester links between the nucleotides with the

Fig. 2.1. The four common ribonucleoside monophosphates.

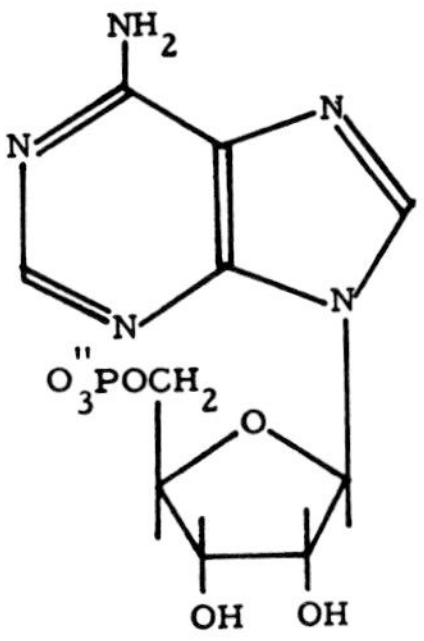

Adenosine-5'-monophosphate
adenylate (AMP)

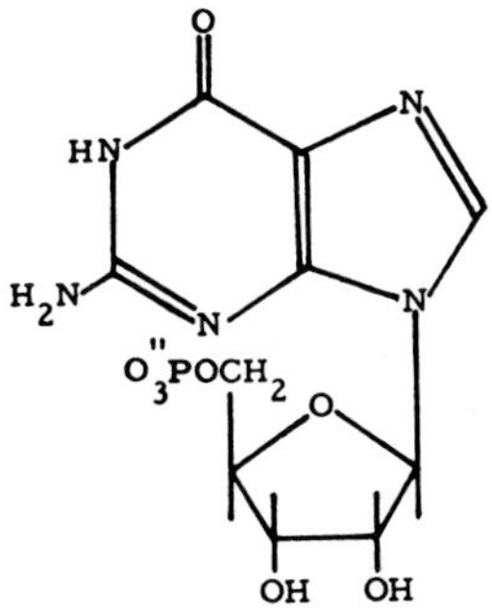

Guanosine-5'-monophosphate
guanylate (GMP)

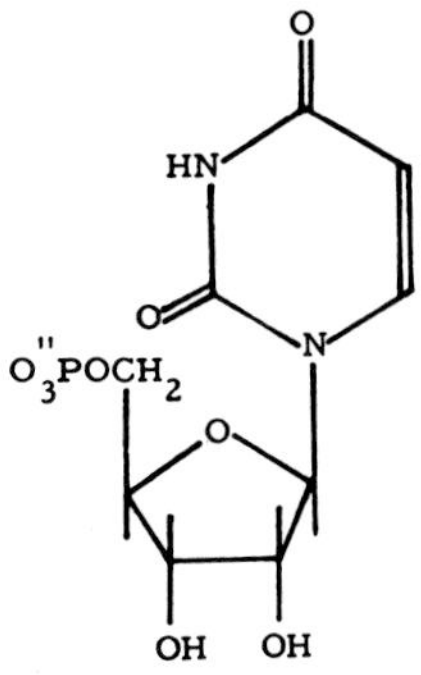

Uridine-5'-monophosphate
uridylate (UMP)

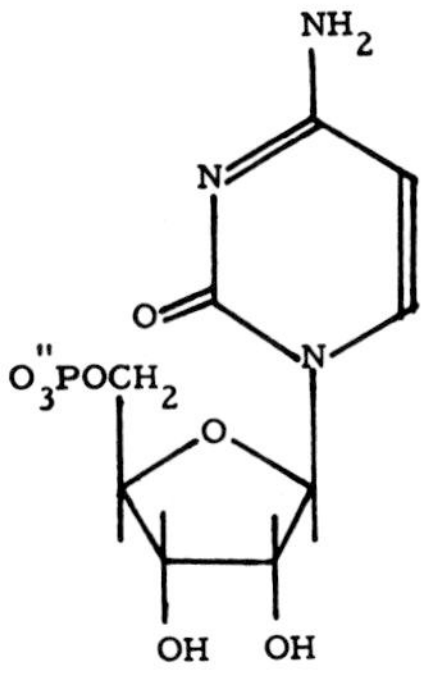

Cytidine-5'-monophosphate
cytidylate (CMP)

release of pyrophosphate, working from the 5′-end of the RNA. *E. coli* (and other prokaryotes) possess just one type of DNA-dependent RNA polymerase, but eukaryotes have at least four different kinds. These are involved in the synthesis of the different kinds of RNA, and recognise different sequences generally situated upstream from the sites at which transcription starts.

The enzyme from *E. coli* exists in two major forms. The core enzyme is a tetramer of composition $\alpha_2\beta\beta'$ while the holoenzyme contains an additional polypeptide called sigma (σ). σ is required for accurate initiation of transcription, and dissociates from the holoenzyme during its passage along the DNA strand that is being transcribed. An additional polypeptide called rho (ρ) is sometimes required for termination of transcription.

Special features of transcription are considered further in Chapters 5 and 7.

2.3 The four major classes of RNA

Many different species of RNA are made in the cell. They are generally grouped into four classes.

1. *Ribosomal RNA* (*rRNA*) is a major structural component of the ribosomes. Three separate molecules are found in prokaryotes, while eukaryotes possess four distinct rRNAs. They are generally designated by their Sedimentation Coefficients (Table 2.1). These rRNAs are transcribed in the form of larger molecules in which the actual rRNA molecules are

Fig. 2.2. A stem and loop structure in RNA. This particular sequence is the terminator structure of the *trp* operon in *E. coli* (Chapter 6.7).

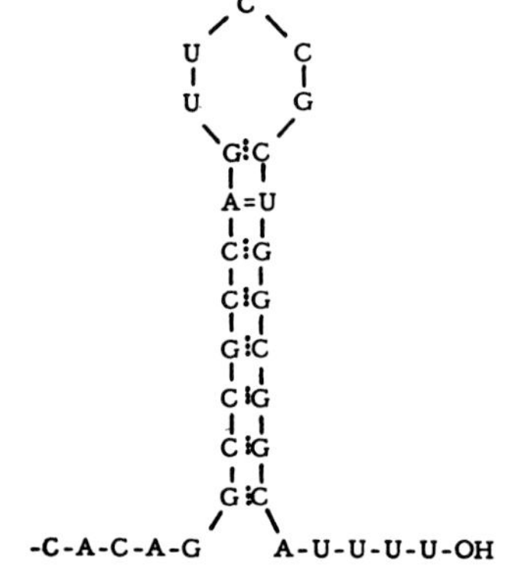

Table 2.1. *Composition and some properties of ribosomes and their subunits*

Subunit	Sedimentation coefficient (svedbergs)	Number of associated proteins	Sedimentation coefficient of RNA (svedbergs)	Molecular weight of RNA × 10^5 (daltons)	Number of bases in RNA
	Prokaryotes				
Large	50	31	23	11	2904
			5	0.4	120
Small	30	21	16	6	1541
	Eukaryotes				
Large	60	45	26* 28†	17	4000–5000
			5.8	0.5	158
			5	0.4	120
Small	40	33	18	7	1800

* Yeast; † Vertebrates.

separated from each other by spacer RNA which has to be removed by endonucleases (Fig. 2.3). In *E. coli* some of the spacers and also the 3′-flanking sequences may contain tRNA molecules. In both prokaryotes and eukaryotes the genes for these pre-rRNAs are repeated. In *E. coli* there are seven copies, while in some amphibia there are up to about 500 copies which are tandemly linked with untranscribed spacer DNA

Fig. 2.3. Eukaryotic precursor rRNA, showing the stages in its cleavage by ribonucleases to the mature molecules.

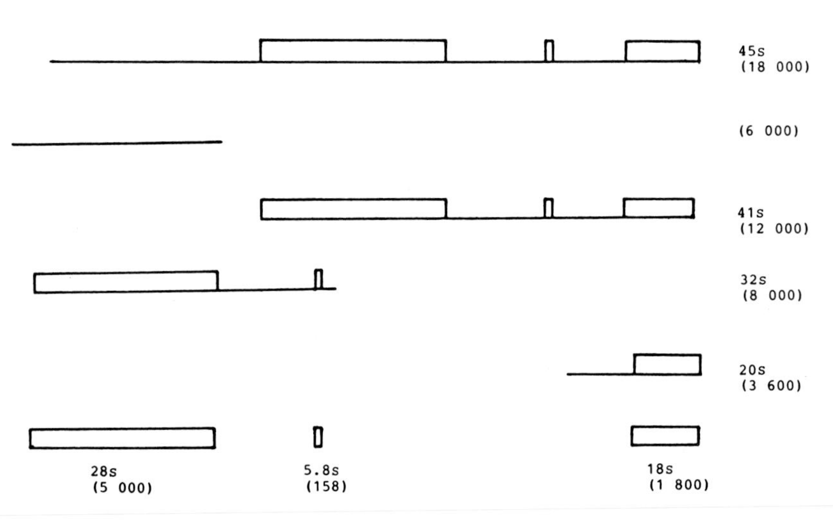

between the rRNA genes. During oogenesis, when the demand for ribosomes increases greatly, these genes are selectively amplified to form about two million copies which are present as extrachromosomal circular DNA.

The *ribosomes* themselves are composed of two subunits, designated large and small. In addition to their RNAs they also contain many proteins. Some details are set out in Table 2.1. The rRNAs in the ribosomes assume characteristic conformations with appreciable hydrogen bonding between bases in the individual molecules.

The structure of prokaryotic ribosomes has been fairly well established by various physical techniques, showing where the proteins are situated in relation to the RNAs. Eukaryotic ribosomes appear to have a similar structure though there are a few extra bulges and lobes in keeping with their larger size.

2. *Transfer RNAs* (*tRNAs*) are a family of smaller RNAs (3–4S), each containing about 80 nucleotides. There are about 40–60 different species in one cell. Like the rRNAs, tRNAs are first made as longer molecules which may have extra nucleotides at one or both ends and also in the interior of the molecule. These are removed in a specific manner by ribonucleases (RNases). The mature molecules contain a number of bases

Fig. 2.4. Some bases found in tRNA as the result of chemical modifications of the four most commonly occurring bases.

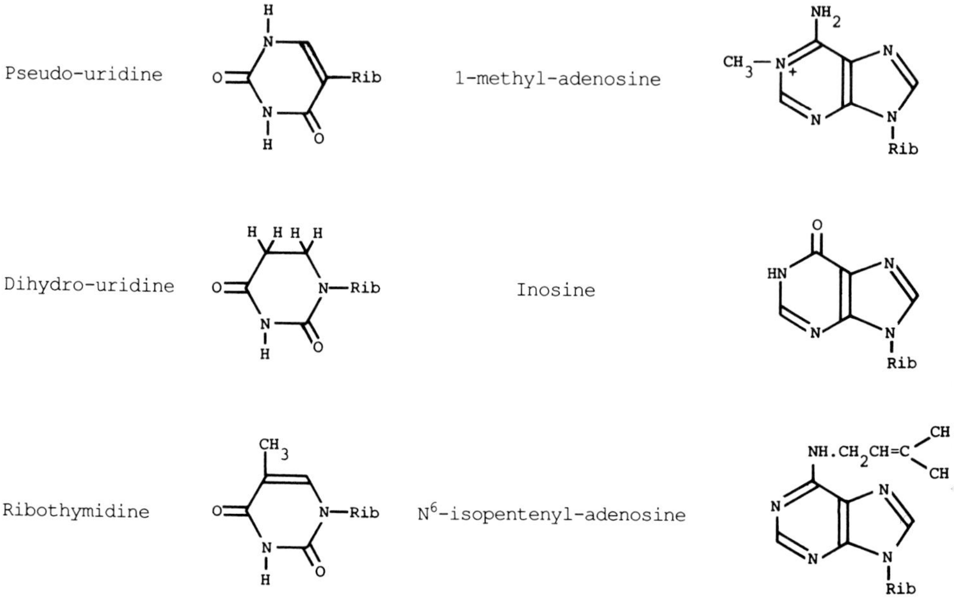

Pseudo-uridine

1-methyl-adenosine

Dihydro-uridine

Inosine

Ribothymidine

N^6-isopentenyl-adenosine

not generally found in other types of RNA (Fig. 2.4), that are produced by enzymic modification of the four types of base in the primary transcript.

The tRNAs possess common structural features with considerable

Fig. 2.5. Above: Yeast tRNA^{Phe} drawn in its clover-leaf form (two-dimensional). D, AC and T are the *d*ihydrouracil, *a*nti*c*odon and *t*hymine loops respectively. The anticodon is the three bases in a horizontal line at the bottom of the figure.
Below: A representation of the same molecule as it appears in three dimensions. (Reproduced from S.-H. Kim, *Adv. Enzymol.* (1978), **46**, 279, by kind permission of John Wiley & Sons, New York, © 1978.)

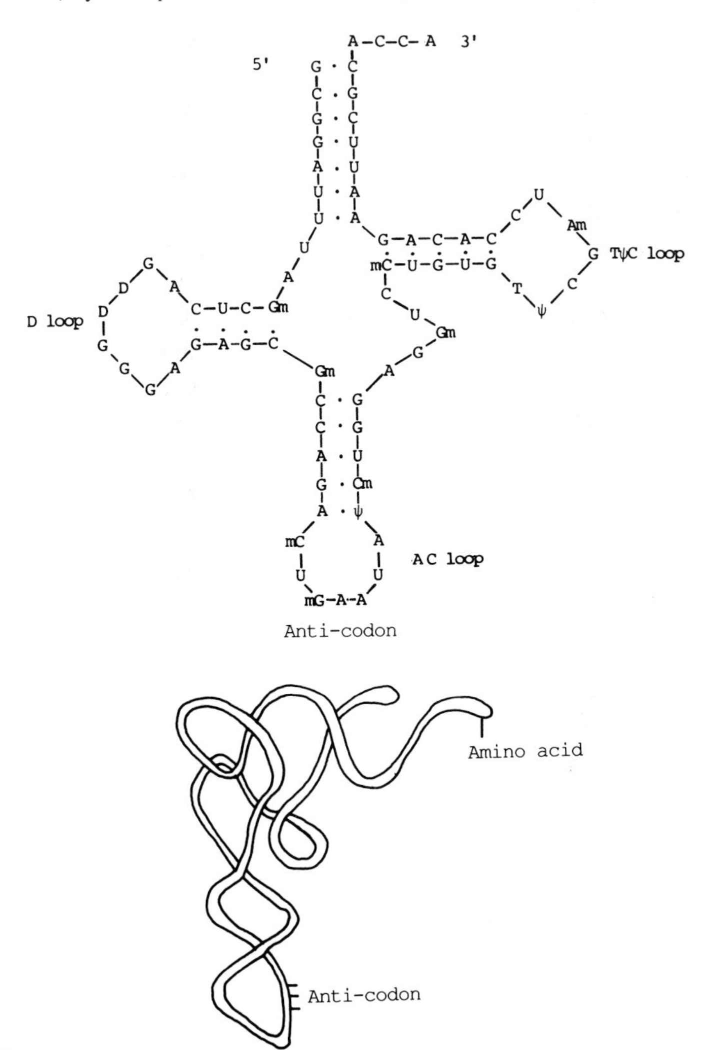

secondary structure, containing four base-paired stems and three or four unpaired loops. They can be represented in two dimensions by a 'clover-leaf' type of structure. In three dimensions they are L-shaped molecules (Fig. 2.5). Each one contains a specific sequence of three bases – the *anticodon* (see Fig. 2.5) – which is situated in the middle of an unpaired loop at one end of the molecule. All tRNA molecules contain the sequence CCA at the 3′-terminus which is at the opposite end to the anticodon. A specific amino acid can be attached to the 2′- or 3′-position of the adenylate residue in this sequence by means of a reaction catalysed by the appropriate amino acyl-tRNA synthetase. When combined with an amino acid a tRNA is said to be 'charged' (Fig. 2.6).

3. *Messenger RNA* (*mRNA*) is the name given to the transcripts that carry the specific information for the sequence of amino acids in proteins. There are a large number of mRNA species of widely varying sizes. They generally possess little secondary structure. In prokaryotes, several mRNAs are commonly transcribed in tandem as one large molecule with very short non-translated spacers between them (see Chapter 6). In eukaryotes primary transcripts synthesised by RNA polymerase are frequently much larger molecules which are processed by excision of parts of the molecule to yield mature mRNAs (see Chapter 7.18). A large proportion of the RNA in the nucleus is in the form of the primary or partly processed transcripts, and is known as *heterogeneous nuclear RNA* (*hnRNA*). mRNAs contain flanking sequences of bases at their 5′- and 3′-ends in addition to the actual coding sequences that specify the amino acid sequence of proteins.

mRNAs are generally not very long-lived, as they are fairly readily degraded by RNases. This is especially true in prokaryotes where their lives are of the order of only a few minutes. In eukaryotes they are normally more stable, with lives ranging from a few minutes to many hours. rRNAs and tRNAs, being much less rapidly degraded, are often referred to as *stable* RNAs.

Fig. 2.6. The charging of a tRNA molecule with its cognate amino acid.

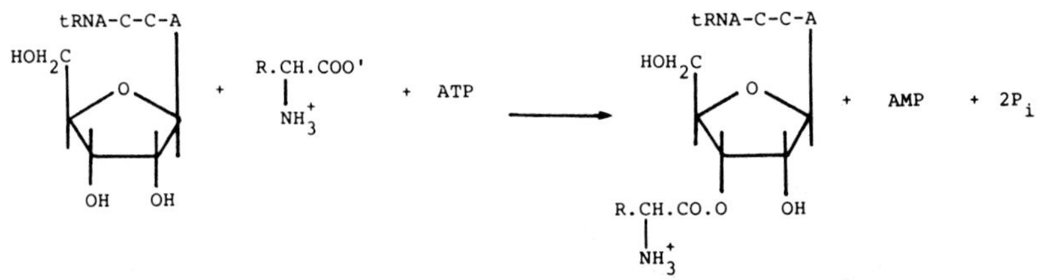

Table 2.2. *Small RNAs*

Vertebrates			Yeast	
Symbol	Number of nucleotides	Copies per cell (thousands)	Symbol	Number of nucleotides
U1	165	1000	snR19	625
U2	188	500	snR20	1125
U3	214	200	snR10 (?)	245
U4	146	200	snR14	160
U5	118	200	snR7	213
U6	108	200	snR6	106
U7	64	30		
U8	140	25		
U11	145	30		

All are found in the nucleoplasm, except for U3, U8 and snR10 which are found in the nucleolus. The yeast homologues are listed on the same line as the vertebrate ones.

4. *Small nuclear RNAs* A number of small RNAs, particularly rich in U residues and generally containing less than 300 nt are found in the nuclei, encoded by multigene families. Collectively they are known as small nuclear RNAs (snRNAs) (Table 2.2). All but U6 contain a 2,2,7,-trimethylguanosine cap – a hypermethylated form of that found in mRNA (Chapter 7.16). They are generally associated with proteins, forming ribonucleoproteins (snRNPs). Some of the proteins are common to all the snRNAs, but others are unique to individual ones. Most of them are precipitable by antibodies recognising the epitope known as Sm that are found in patients with the autoimmune disease systemic lupus erythematosus. This epitope is part of a protein that binds to the sequence $RA(U)_nGR$ on the RNA, or a close relative of this. The sequences of the snRNAs are highly conserved between insects, vertebrates and plants.

The snRNPs function in various ways in processing the primary transcripts of rRNAs and pre-mRNAs (Chapter 7.6 and 7.18). U1, U2, U4, U5, and U6 are found in high abundance (10^5–10^6 molecules per cell), consistent with their role in the removal of introns (Chapter 7.18), but most of the others are less abundant by two or more orders of magnitude. So far, U7 and U11 have been definitely implicated in 3′-end processing. Since, on average, eukaryotic genes contain about ten introns, it is reasonable that the snRNAs involved in their removal from pre-mRNAs should be much more abundant than those involved in 3′-end processing.

Yeast snRNAs have been analysed and described separately. None are as abundant as those in other eukaryotes. Several are dispensable for apparently normal growth and function. Some are used for splicing pre-mRNAs in a manner analogous to U1, U2, U4, U5 and U6. Yeast snR20 is very much larger than the other snRNAs, but at its 5′-end there is a sequence of 180 nt that is very similar to that of U2. The comparatively low abundance of these yeast snRNAs is in accord with the rare occurrence of introns in yeast pre-mRNAs (Chapter 7.17). Yeast snR10 has been implicated in the maturation of rRNAs from their longer precursors, and it is likely that vertebrate U3 and U8 are also involved in this process, since both are found in the nucleolus where this activity occurs. U3 is not precipitable by the anti-Sm antibodies, and is associated with a set of proteins different from those of the other abundant snRNAs.

The transcription of the snRNA genes is extremely efficient since over a million snRNA molecules are needed for each cell in each generation. After their synthesis these RNAs are rapidly exported to the cytosol where they assemble into ribonucleoprotein particles that are transported back into the nucleus.

Several small RNAs occur outside the nucleus. The best characterised is the 7SL-RNA that is part of the signal recognition particle, involved in cleavage of signal peptides from precursors to secreted proteins (Chapter 7.19). There are also several small RNAs in the cytosol (scRNAs) occurring as ribonucleoprotein complexes. Their functions are not definitely known, but they may play a role in the translation of mRNAs at the ribosomes.

2.4 The genetic code

The information encoded in the DNA that directs the synthesis of specific proteins is in the form of a nearly universal genetic code in which a sequence of three bases in a nucleic acid codes for a single amino acid residue in a protein. Such a triplet of three bases is known as a *codon*, and 64 codons can arise by permutating the four bases in triplets. Since only 20 different amino acids are incorporated into proteins, either many codons are not used, or some amino acids can be coded for by more than one codon. The latter is the true situation, and the code is therefore said to be *degenerate*. Three codons are also used for stop signals (*termination codons*). These are sometimes known as nonsense codons, and have been given the fanciful names amber, ochre and opal.

Codons are carried on mRNAs and are therefore also found on the DNA strand that is not transcribed (with T replacing U). The complementary triplets are known as anti-codons, and are found on the transcribed strand of DNA (known as the sense strand), and also in tRNAs. When writing the DNA sequence of protein-coding genes, it is

Table 2.3. *The genetic code*

First base (5′-end)	Second base U	C	A	G	Third base (3′-end)
U	Phe	Ser	Tyr	Cys	U
	Phe	Ser	Tyr	Cys	C
	Leu	Ser	Ter	Ter	A
	Leu	Ser	Ter	Trp	G
C	Leu	Pro	His	Arg	U
	Leu	Pro	His	Arg	C
	Leu	Pro	Gln	Arg	A
	Leu	Pro	Gln	Arg	G
A	Ile	Thr	Asn	Ser	U
	Ile	Thr	Asn	Ser	C
	Ile	Thr	Lys	Arg	A
	Met	Thr	Lys	Arg	G
G	Val	Ala	Asp	Gly	U
	Val	Ala	Asp	Gly	C
	Val	Ala	Glu	Gly	A
	Val	Ala	Glu	Gly	G

Amino acids are designated by their standard three-letter abbreviations. Ter is a termination codon.

usual to write the sequence of the non-coding strand, as this will contain codons in the correct order so that the amino acid sequence of the protein can be read off directly.

Inspection of the genetic code (Table 2.3) shows that the first two bases of the codon have the greatest effect in specifying the amino acid to be incorporated into a protein at a given position. One consequence of the degeneracy of the code is that many point mutations are ‘silent’, i.e. not expressed since, for example, mutation of the codon UUC to UUU still leads to the incorporation of phenylalanine. It is also found that the 5′-base of the anticodon in tRNA does not always pair exclusively with the complementary 3′- one on a codon. The base hypoxanthine is sometimes found in the 5′-position of anticodons, and this can pair with C, U, or A in the 3′-position of the codon. This lack of specificity was first proposed by Crick in what is known as the *wobble hypothesis.* Consequently some species of tRNA can recognise more than one codon, leading to economy in the production of these molecules so that less than the theoretical number of 61 tRNAs are needed to read all the codons on an mRNA.

The codon AUG for methionine is very nearly always used as an initiation codon for the first amino acid to be incorporated into a polypeptide chain. Successive triplets are read to give the specified protein sequence until a termination codon is reached. Thus, point mutations in which a single base is changed will lead to the incorporation of a single different amino acid (unless they are silent) or, in rarer cases, to the premature termination of a polypeptide chain or to the reading through of a termination codon. However, in mutations resulting in deletions or insertions of one or two bases, the reading frame will be thrown out of register from that point onwards, so a completely different protein will be synthesised (Fig. 2.7).

Very exceptionally, in some bacterial viruses with small genomes, two or even three reading frames are used simultaneously so that portions of a long mRNA can code for parts of two or three proteins (Fig. 2.8). Intuitively, it seems surprising that two functional proteins can arise in this way, but it obviously leads to significant economy in the amount of coding DNA that is required.

Iso-accepting tRNAs (different tRNAs that can be charged with the same amino acid although they bear different anticodons) are not all present in a cell in the same amount, and the less-abundant ones are used less frequently. Codon usage is not random. In both yeast and *E. coli*, mRNAs that code for proteins of low abundance contain a higher than usual proportion of 'rare' codons. Thus, to quote a fairly extreme case, the yeast genes coding for three proteins which are abundantly expressed (phosphoglyceraldehyde dehydrogenase, alcohol dehydrogenase, enolase) use only 27 out of the possible 61 codons specifying amino acid residues more than 98% of the time.

Fig. 2.7. A single nucleotide deletion in the gene for Haemoglobin$_{\mathrm{Wayne}}$ throws the reading frame out of register, and leads to the synthesis of a longer α-chain. The codons of the mRNAs of the normal and mutated α chains are shown, together with the amino acids they specify, written in the standard three-letter code. The deleted nucleotide is boxed.

```
Hb        Thr Ser Lys Tyr Arg Ter
          ACN UC[U]AAA UAC CGU UAA GCU GGA GCC UCG GUA G

HbWayne   ACN UCA AAU ACC GUU AAG CUG GAG CCU CGG UAG
          Thr Ser Asn Thr Val Lys Leu Glu Pro Arg Ter
```

Fig. 2.8. Overlapping genes in the bacteriophage X174. The beginning and end of the coding portion of the gene for protein E, and the end of the coding portion of the gene for protein D are shown, together with the amino acids specified, written in the three-letter code.

```
 Met Val Arg Tyr                          Ter
 ATG GTA CGC TGG - - - protein E - - - G TGA TGT AA

TAT GGT ACG CTG G - - - protein D - - -GTG ATG TAA
Tyr Gly Thr Leu                        Val Met Ter
```

2.5 Translation is a later stage of gene expression

Gene expression involves translation of the triplet base code carried on mRNA into the sequence of amino acids in a protein. It is customary to divide it into three phases occurring sequentially.

The *initiation codon* is very nearly always AUG. This is the only codon for methionine, but there are two different tRNAs each containing the anticodon for this (CAU). In practice, the one that is always used for initiation in prokaryotes ($tRNA_i^{Met}$) is charged with N-formyl-methionine (Fig. 2.9). A different tRNA ($tRNA_m^{Met}$) is used for inserting methionine into internal positions in a peptide chain. In eukaryotes there are again two tRNAs for methionine, but the methionine that is used for initiation is not formylated. Since the majority of mature proteins do not contain methionine as the N-terminal amino acid, the initiating methionine residue is often cleaved from the protein post-translationally.

Protein synthesis in prokaryotes starts with charged $tRNA_i^{Met}$ binding to the initiation codon on the mRNA in the presence of soluble protein initiation factors. The mRNA binds by base pairing of the complementary Shine–Delgarno sequence (Chapter 5.2) in its 5′-flanking region to the 16S RNA in the small subunit of the ribosome. Finally the large ribosomal subunit is bound, GTP is hydrolyed to GDP and Pi and the initiation factors dissociate from the complex (Fig. 2.10).

Initiation of translation in eukaryotes starts with the binding of a trimeric cap binding protein to the 7-methyl-guanyl residue at the 5′-end of the mRNA (Chapter 7.16). One of the subunits of this protein binds to the RNA with concomitant hydrolysis of ATP to unwind any secondary structure that may have formed in the 5′-untranslated sequence. This complex then associates sequentially with the 40S and 60S ribosomal subunits. Finally, the mRNA migrates so that the complex is positioned correctly for interacting with the charged $tRNA_i^{Met}$. These reactions require a number of soluble initation factors, and GTP is hydrolysed to GDP and Pi.

Fig. 2.9. The charging of $tRNA_i^{Met}$ and formylation of the methionyl residue prior to initiation of protein synthesis in prokaryotes. PP_i = inorganic pyrophosphate; THF = tetrahydrofolate.

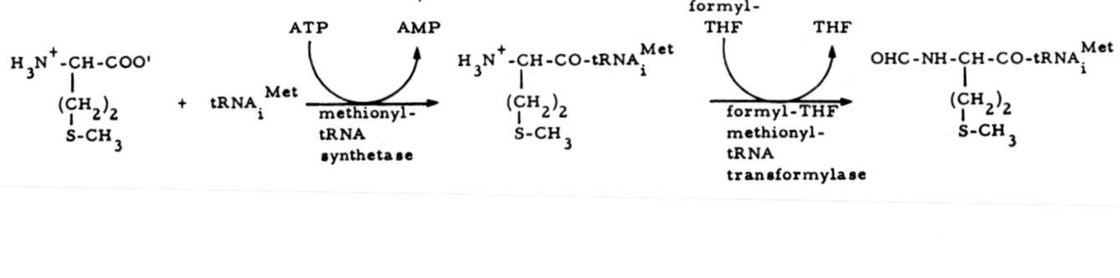

H_3N^+-CH-COO' | $(CH_2)_2$ | $S-CH_3$ + $tRNA_i^{Met}$ —(ATP → PP_i + AMP; methionyl-tRNA synthetase)→ H_3N^+-CH-CO-$tRNA_i^{Met}$ | $(CH_2)_2$ | $S-CH_3$ —(formyl-THF → THF; formyl-THF methionyl-tRNA transformylase)→ OHC-NH-CH-CO-$tRNA_i^{Met}$ | $(CH_2)_2$ | $S-CH_3$

The nucleotide context in which the initiating methionine codon appears is important, since in the majority of eukaryotic mRNAs there is a consensus sequence GCCRCCAUGG around the site where translation commences. Occasionally there are other AUG sequences in the mRNA upstream from this site, but they are generally embedded in different sequence and are not recognised as suitable sites for initiation of translation.

In both cases the initiating charged tRNA is bound at a site known as the P (peptide) site, where the actual peptide links are formed.

Elongation is the second phase of translation (Fig. 2.11) when the next available codon on the mRNA pairs with a new charged tRNA molecule. This directs its amino acid residue to the A site on the ribosome next to the initiator methionine, which is then transferred onto the amino group of the incoming amino acyl residue. Soluble protein elongation

Fig. 2.10. Initiation of protein synthesis in eukaryotes. Initiation in prokaryotes is very similar in outline, but uses formyl-methionyl-$tRNA_i^{Met}$ and the ribosomal subunits are 30s and 50s with the formation of a 70s complex. IFs = initiation factors.

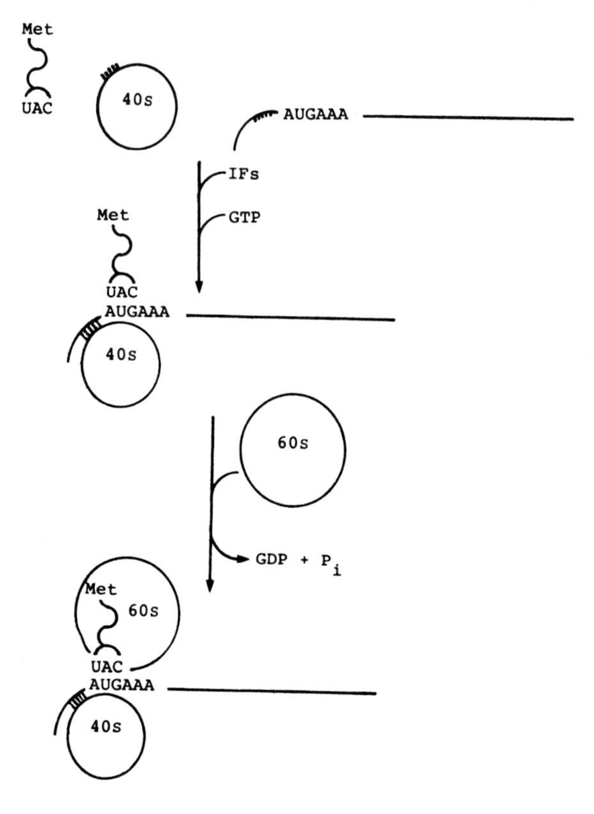

factors are required for this process. The uncharged $tRNA^{Met}$ dissociates from the ribosome and the charged tRNA (now bearing a dipeptide) and the mRNA move relative to the ribosome. Another elongation factor is used in this process and another molecule of GTP is hydrolysed. This brings the dipeptidyl tRNA to the P site so that a new charged tRNA can

Fig. 2.11. Elongation step of protein synthesis. Note the use of *e*longation *f*actors (EFs), and the hydrolysis of GTP to provide energy for the reaction to proceed.

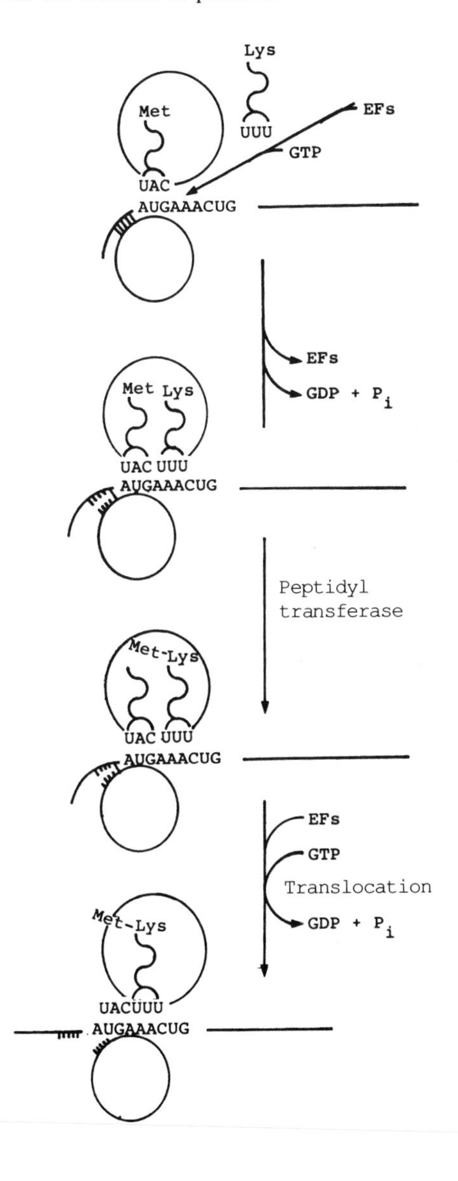

enter at the A site by base pairing with the next codon on the mRNA. The polypeptide chain is built up sequentially by many repetitions of this process. Amino acid residues are incorporated into growing polypeptide chains at a rate of about 100 residues a second.

The termination phase of translation occurs when a termination codon is reached. Now there will be no corresponding tRNA and the peptide chain that has been built up is hydrolysed off the last tRNA which has been used. Soluble protein termination factors are involved which associate briefly with the ribosome. With the dissociation of the last, now

Fig. 2.12. Electron micrograph of polysomes. This is actually from a bacterial preparation. (Reproduced from O. L. Miller Jr, B. A. Hamkalo & C. A. Thomas Jr, *Science* (1970), **169**, 392–5. © 1970 by the AAAS.)

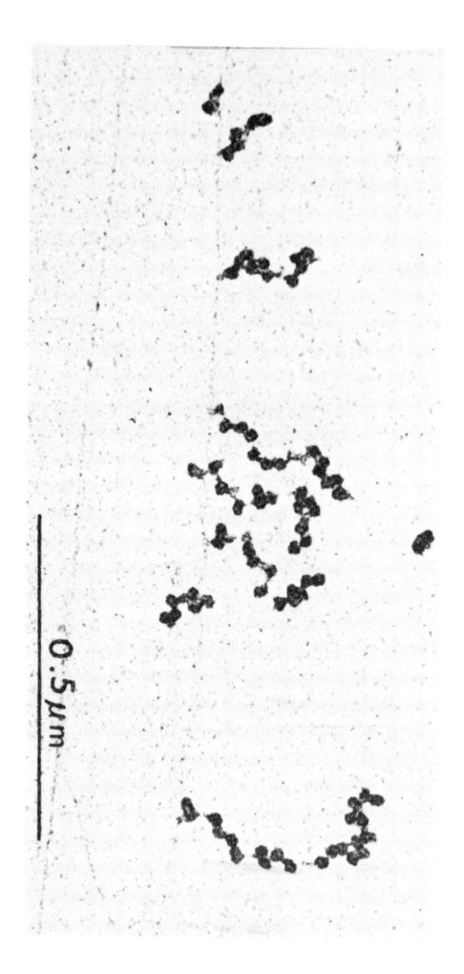

uncharged, tRNA from the complex the ribosomal subunits also dissociate from the mRNA and from each other to join the pool of subunits which are available for reassociation in the initiation complex.

Several ribosomes are commonly found spaced out along an mRNA molecule separated from each other by about 80–100 bases. The mRNA coding for globin, a protein of about 145 amino acid residues, has a coding sequence of about 450 bases and could accommodate up to about five ribosomes. Such a structure is called *a polysome*. Fig. 2.12 shows several polysomes formed on the mRNA as it is transcribed from a strand of DNA in a bacterial preparation.

Thus, a single very complex process is used for the synthesis of proteins under the direction of specific mRNAs coding for particular proteins. Fuller details can be found in any good up-to-date textbook of biochemistry.

3

Methodology

3.1 Introduction

The recent explosive growth in knowledge of the structure and function of the genome has only been possible because of the development of suitable techniques. A proper appreciation of the methods that are widely used and of the jargon that goes with them should lead to a better understanding of the work that has been done.

Much of this work has centred on *genetic engineering* in which fragments of DNA are excised from their natural chromosomal sites and incorporated into larger pieces of DNA that are capable of autonomous replication in cells of various kinds. Under suitable conditions this allows amplification of the original DNA fragments so that they can be produced in large enough quantities and in a pure enough state for further study.

DNA can be obtained for these purposes in four ways:

1. By stirring a suspension of DNA at very high speed at 0 °C to yield randomly sheared fragments.
2. By enzymic synthesis on an RNA template using the enzyme reverse transcriptase (Chapter 3.6) to produce *complementary DNA* (*cDNA*).
3. By the action of restriction endonucleases (Chapter 3.9) for hydrolytic cleavage at specific sites.
4. By chemical synthesis that can be automated to produce oligonucleotides 50–100 nt long in a reasonable time. This method will only make fragments smaller than those produced by the other methods, some of which can yield pieces of DNA up to at least 60000 bp long.

DNA made in one of these ways is ligated to the DNA of an independently replicating vector that will grow when introduced into appropriate cells (Chapter 4) to yield clones of cells infected with a particular type of DNA. Hence the technique is often referred to as cloning of DNA. The artificially produced DNA may be called *recombinant DNA* since DNA will have been combined from two or more sources for its production.

Fears were expressed that bacteria that had incorporated plasmids containing eukaryotic DNA might escape from laboratories and could be hazardous to health since they would be expressing eukaryotic (including human) genes in the wrong places. These fears have proved to be unfounded, largely because the recombinant vectors are grown in mutant strains of bacteria that require nutrients which are not normally available outside a strictly controlled laboratory environment.

3.2 mRNA isolation

mRNAs are most easily isolated and purified from eukaryotic cells, particularly in tissues that synthesise only a very limited number of proteins with one main product, such as reticulocytes synthesising predominantly haemoglobin and plasmacytomas synthesising predominantly immunoglobulins. After separation, the microsomal fraction is freed of proteins and lipids by suitable extractions, leaving the RNA. Most eukaryotic mRNAs possess a long tail of poly-A residues up to 200 nt long, so the crude RNA is passed down a column containing either poly-U or poly-T linked to a solid support. tRNA and rRNA pass through but the poly-A-containing RNA is bound and can be eluted by raising the ionic strength of the eluant. Further purification can be effected by electrophoresis in favourable cases.

When a tissue produces many different proteins it is more difficult to separate one particular mRNA. However, this can generally be achieved by immunoprecipitation. If antibodies to the pure protein specified by the desired mRNA are added to the crude microsomal fraction only those polysomes synthesising that protein will be precipitated. The mRNA can then be separated from the other polysomal constituents as described above.

Purification of prokaryotic mRNAs is more difficult since they contain no distinctive features like a poly-A tail. They are also much more readily hydrolysed by endogenous RNases *in vivo*. Isolation of prokaryotic genes is generally accomplished by hydrolysing the total DNA with restriction endonucleases and isolating fragments in the 8–18 kbp size range. These can be cloned into suitable vectors from which individual cells are isolated and grown on. Those producing the product of the desired mRNA are selected by means of a sensitive test for it.

3.3 Polymerase chain reaction

A very elegant method has been developed to synthesise relatively large quantities of a particular deoxynucleotide sequence. The double-stranded DNA containing the desired sequence is prepared, generally by cutting genomic DNA with suitable restriction endonucleases. Then two oligodeoxyribonucleotides are synthesised – one exactly complementary to about 20 nucleotides at the beginning of the required sequence, and the other, of a similar size, exactly complementary to about 20 nucleotides on the opposite strand at the other end of this sequence. A large molar excess of these oligonucleotides is mixed with the target DNA and heated briefly to 90 °C to denature the double-stranded DNA. On cooling to 75 °C, the denatured strands will associate with some of the oligonucleotides. These are added to an even larger excess of the four deoxynucleoside triphosphates plus DNA polymerase from the thermophilic bacterium *Thermus aquaticum* and any necessary cofactors. After about two minutes, appreciable quantities of each strand of the target DNA will have been

Fig. 3.1. Polymerase chain reaction. Top: double-stranded DNA to be amplified with synthetic primer at each end. Middle: the primers are annealed to each strand. Bottom: newly synthesised DNA strands that have based paired with the original complementary strands.

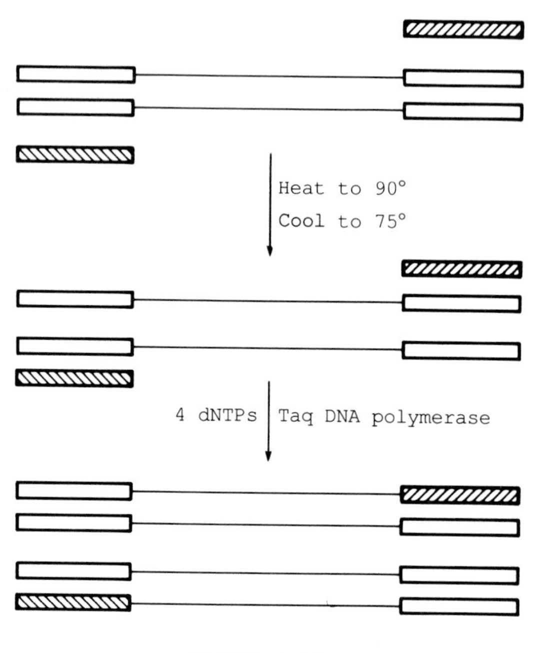

synthesised, beginning exactly at the desired 5′-ends. The whole process of heating and cooling to 75 °C is now repeated up to about 60 times. As each cycle of the process produces more of the required DNA, the synthesis of this fragment proceeds at an exponential rate. The enzyme from the thermophilic bacterium is not denatured and remains active despite the high temperatures required for constantly breaking and reforming the DNA duplexes. With increased lengths of DNA to be synthesised, it is possible that some undesired copies may be obtained, but the product can be purified by electrophoresis if need be (Fig. 3.1).

3.4 Electrophoresis of nucleic acids

The size of both RNA and DNA molecules can be determined by measuring their electrophoretic mobility in gels of polyacrylamide or agarose. The mobility of any sample depends on its size and also on the composition of the gel. The higher the concentration and degree of cross-linking of a polyacrylamide gel, the slower the nucleic acid molecule will migrate. It is not generally possible to separate fragments outside a fairly narrow range of sizes on one gel. Samples containing fragments spanning a wider range of sizes may be separated on two (or even more) gels made from different concentrations of polyacrylamide.

Single-stranded molecules will migrate differently if allowed to form secondary structures, so gels for analysing these molecule contain denaturants such as high concentrations of urea or formamide, and the DNA samples can be denatured first by treatment with glyoxal or dimethyl sulphoxide. In a given gel, the rate of migration of a nucleic acid is proportional to the logarithm of its chain length. A series of standards are usually run in parallel with the experimental samples. These are prepared by digestion of a plasmid or phage DNA with a restriction enzyme to generate fragments of known lengths. Nucleic acids are detected on gels by staining with ethidium bromide which has an intense fluorescence excited by ultraviolet radiation when it is complexed with nucleic acids.

The usefulness of electrophoresis has been greatly extended by the development of *Pulsed Field Gel Electrophoresis* (PFGE). Ordinary electrophoresis is limited to the separation of polynucleotides of less than about 20 kbp in length that can migrate reasonably freely through the pores in the gel matrix. As the length increases, the separation becomes less and less good. In PFGE, the electric current is delivered in short pulses at an angle to each other. Smaller molecules orient themselves more rapidly and so will move through the gel faster than larger ones. Major variables in this technique are the angle between the fields generated by

Fig. 3.2. Pulsed Field Gel Electrophoresis. Separation of yeast chromosomes with Pulsaphor™ Electrophoresis Unit and Pulsaphor Plus Controller Unit (Pharmacia LKB). 15 of the 16 chromosomes are separated, with chromosomes XV and VII forming the bright band that is fourth from the top. The sizes of the chromosomes range from 225 to 1990 kbp. The DNA was visualised by staining with ethidium bromide after electrophoresis and viewed in ultra-violet light. (Reproduced by kind permission of Dr Tom Tyre of Technical Services, Pharmacia P-L Biochemicals.)

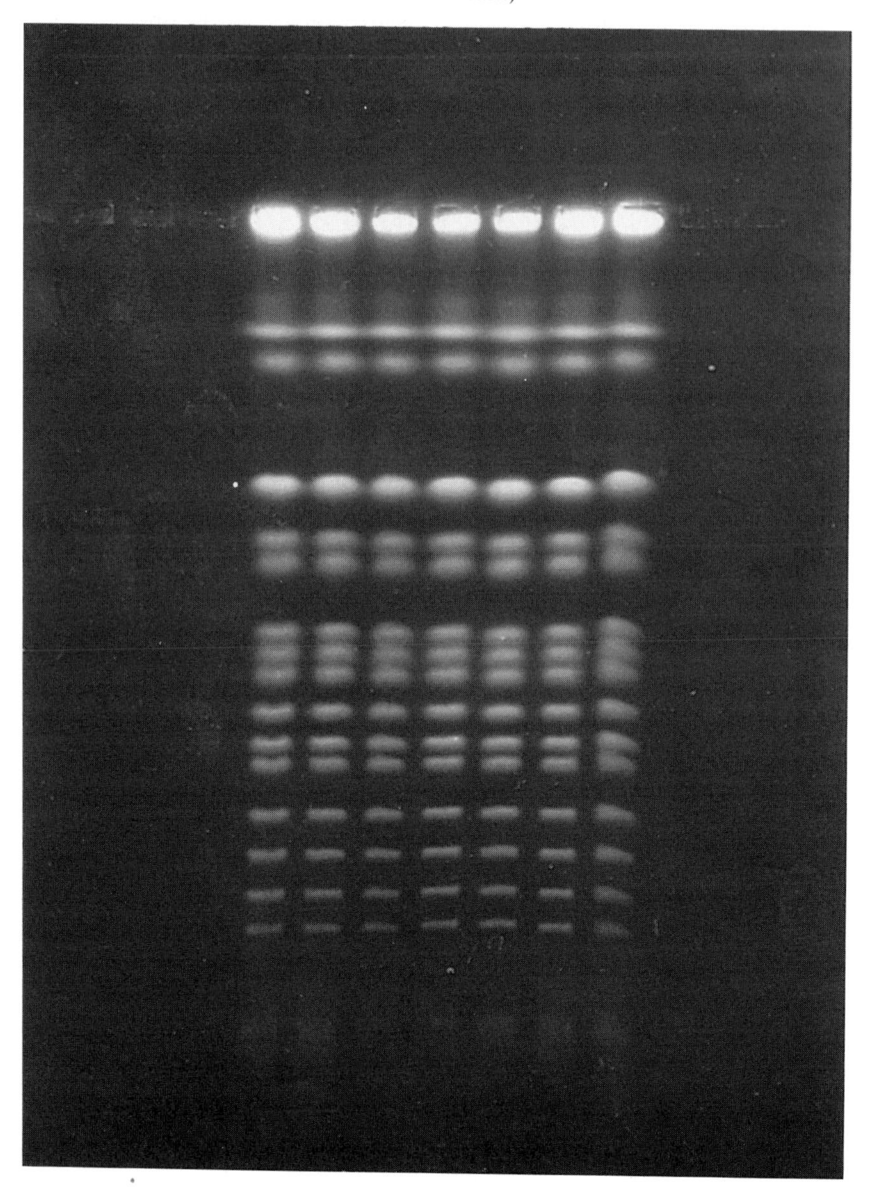

successive pulses and also the length of the pulses, from seconds for smaller DNA molecules up to an hour for the largest molecules. The best conditions for any separation are worked out empirically or on the basis of previous experience.

Molecules up to 10 Mbp long, such as whole chromosomes of some organisms like fungi, can be separated with this technique. For investigating genomic DNA of higher plants and animals with longer chromosomes, rare cutting restriction endonucleases are used to generate fragments 10–100 kbp long. The recognition sequences of Not 1 and Sfi 1 are eight nucleotides long, which are longer than those of other such enzymes (Table 3.3). On average, they will only cleave DNA once every 64000 nt. Figure 3.2 shows a typical separation made with this technique.

Another widely used electrophoretic technique is a *gel mobility shift* assay to detect the binding of proteins to DNA. If a protein binds to a fragment of DNA, its electrophoretic mobility decreases because the complex has a larger size. This can be readily detected after electrophoresis by autoradiography of the gel if the DNA fragment is suitably labelled (Fig. 3.3).

3.5 Footprinting

A protein that binds strongly to a specific base sequence in DNA will protect that sequence from digestion with a DNase. This can be visualised by comparing the electrophoretic pattern of the protected DNA and the naked DNA after partial digestion with a DNase. Fragments generated by hydrolysis at the nucleotides in the binding site on the DNA are absent from the protected DNA, though present in the naked DNA that is run as a control. Such a length of protected DNA is known as a *footprint*. At the same time the binding of a substance to DNA may cause conformational changes, increasing the accessibility to DNase of adjacent nucleotides so that hydrolysis at these points is favoured and fragments terminating in this region will be more abundant than in the control partial digests (Fig. 3.4).

Since the relatively large DNase molecule may not be able to penetrate to residues immediately adjacent to the sites where a protein has bound, smaller reagents can be used to pinpoint more accurately the actual binding sites. A particularly useful reagent on account of its very small size and high reactivity is the hydroxyl radical, generated by reaction of Fe II with hydrogen peroxide in the presence of EDTA and ascorbate. Figure 3.4 compares the footprints obtained with DNase and hydroxyl radicals.

The precise guanine residues that interact with a protein can be determined by a rather similar technique. Methylation of guanine residues is the first step in the Maxam and Gilbert method for determining their position in a sequence (Chapter 3.13). If this set of reactions is carried out in the presence and absence of the protein, two sequence ladders will show which guanine residues have been protected from methylation and therefore make close contacts with the protein, so that they cannot be methylated (Fig. 3.5).

3.6 Reverse transcriptase

This is an enzyme that was originally detected in viruses whose genome consists of RNA, rather than DNA. It catalyses the synthesis of

Fig. 3.3. Gel Mobility Shift. A 37-base oligonucleotide containing the binding site for Epstein-Barr Virus Nuclear Antigen was synthesised and labelled with ^{32}P, and incubated with or without an extract containing the antigen and competing oligonucleotides before electrophoresis using the Pharmacia PhastSystem. Lane 1: no extract; Lane 2: with extract; Lanes 3 and 4: with extract plus 13-fold and 25-fold excess of unlabelled oligonucleotide; Lane 5: as Lane 2 plus 38-fold excess of an unrelated oligonucleotide. B = bound oligonucleotide; F = free oligonucleotide. (Reproduced with permission of Dr Ramanujam. Biotechnics 1989 p. 201. © Eaton Publishing Co. Natich. Mass.

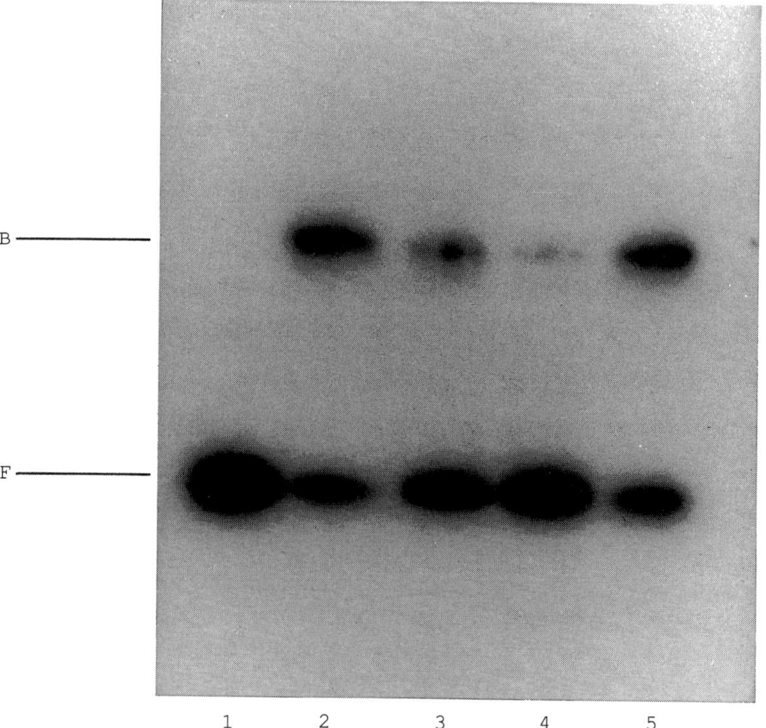

a single strand of DNA complementary to an RNA strand, transcribing information in a direction opposite to that employed in the transcription of DNA by RNA polymerase – hence the name reverse transcriptase.

Complementary DNAs (cDNAs) can be made by this enzyme using RNA molecules as templates, producing single-stranded DNAs that can be converted into double-stranded DNA by the action of DNA polymerase. Amplification of these cDNAs in suitable vectors (Chapter 4) produces large quantities of DNA for further study. Preparation and sequencing of cDNAs is widely used to determine the sequence of proteins since DNAs are much easier to sequence than proteins.

Fig. 3.4. Footprints with DNase digests and hydroxyl radicals. End-labelled *Xenopus* 5S RNA gene digested in the presence (+) and absence (−) of TF IIIA (Chapter 7.7). Lanes 1–4: coding strand; Lanes 5–8: non-coding strand. Note the enhancement of digestion by DNase at certain positions in the presence of TF IIIA. (Reprinted with permission from T. D. Tullius *et al.*, *Methods in Enzymol.* (1987), **155**, 556. © 1987 John Wiley & Sons Incorporated.)

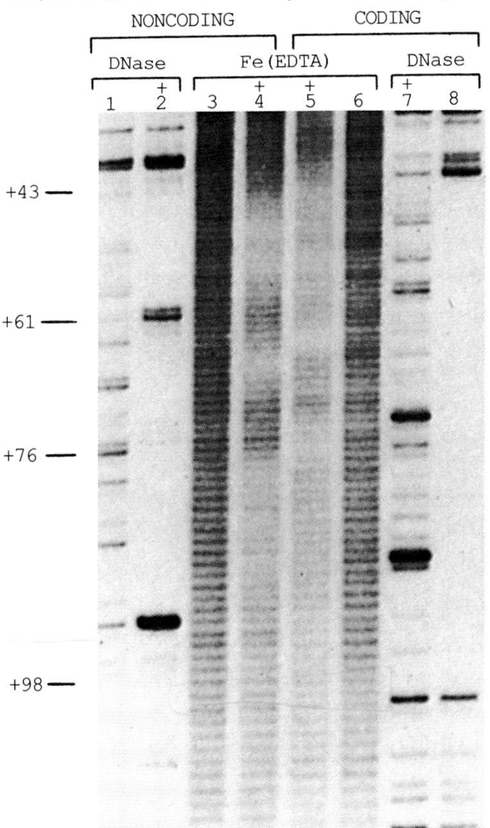

3.7 Site-directed mutagenesis

The need sometimes arises to make proteins with specific substitutions at a defined position in their sequence. This is particularly applicable in the study of the roles of particular residues in the binding or catalytic sites of enzymes. If the sequence of the cDNA for the protein is available, it is possible to alter a codon for any desired amino acid so as to substitute another in its place. In practice, part of the cDNA containing the codon that is to be changed is cut out with a restriction endonuclease and replaced with a synthetic oligonucleotide containing the desired

Fig. 3.5. Methylation protection and DNase footprints of glucocorticoid responsive element (GRE) of human papilloma virus-16. Lanes 1 and 2: DNase footprints in absence (−) and presence (+) of glucocorticoid receptor protein; Lane 3: Maxam & Gilbert sequencing reaction for A and G; Lanes 4 and 5: Maxam & Gilbert reaction for G after treatment with dimethyl sulphoxide in absence (−) and presence (+) of glucocorticoid receptor protein. (Reprinted with permission from B. Gloss *et al.*, *EMBO J.* (1987), **6**, 3737. © 1987 IRL Press Limited)

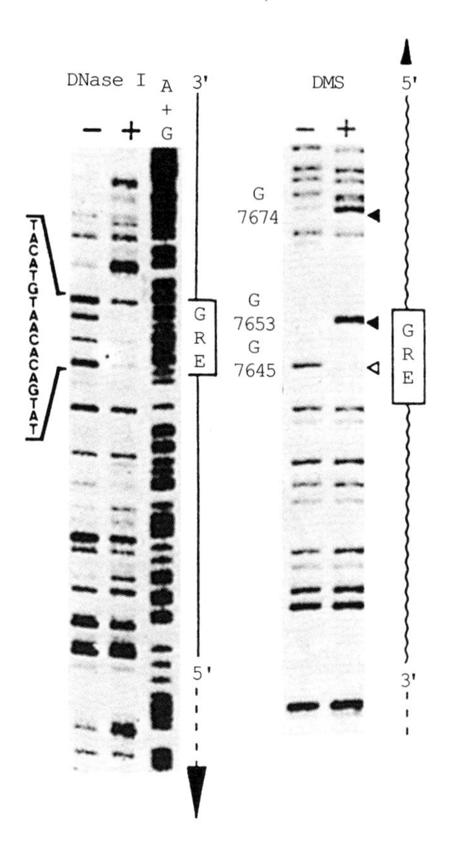

Table 3.1. *General properties of nucleases*

Type of enzyme	Particular characteristics
Ribonuclease	Hydrolyses RNA
Deoxyribonuclease	Hydrolyses DNA
Endonucleases	Attack internal phosphodiester bonds
Exonucleases	Attack the terminal phosphodiester bond, either at the 5′- or 3′-end of the chain

Table 3.2 *Base specificities of ribonucleases*

Name	Site of cleavage	Source
A1	Up N and Cp N	Mammalian pancreas
Phy M	Up N and Ap N	*Physarum polycephalum*
T1	Gp N	*Aspergillus oryzae*
U2	Ap N	*Ustilago sphaerogena*

These enzymes are all endonucleases, yielding 3′-phosphate and 5′-hydroxyl groups on the cleavage products.

codon. The engineered cDNA can then be expressed in a suitable vector and the protein purified for study.

3.8 Nucleases

Nucleases hydrolyse the phosphodiester bonds in nucleic acids and are important in processing RNA and DNA in cells. Table 3.1 sets out some general properties of these enzymes.

RNases that only hydrolyse phosphodiester bonds next to a particular nucleotide (Table 3.2) are useful in elucidating the structure of RNA. Others that recognise tertiary structures are used in processing primary transcripts. RNase III of *E. coli* recognises and cleaves the double-stranded stem of the precursor to rRNAs (Chapter 5.3).

RNase P generates the 5′-terminus of mature tRNAs by endonucleolytic hydrolysis of precursors. It consists of a small protein plus a very much larger RNA, known as M1 RNA. In the presence of unphysiologically high concentrations of Mg^{2+}, the RNA alone, without the protein, can catalyse the processing of tRNA precursors. However, the addition of the protein increases the rate of this reaction. The enzyme is widely distributed in prokaryotes, and there is a similar one in eukaryotes.

The endonuclease RNase H acts on hybrid RNA–DNA chains, trimming off ribonucleotides that have been covalently linked to DNA after their use as primers for DNA replication (Chapter 5.1).

Table 3.3. *Some restriction endonucleases*

Symbol	Organism and strain	Sequence recognised
EcoRI	*Escherichia coli*	G|AATTC CTTAA|G
Hpa I	*Haemophilus parainfluenzae*	GTT|AAC CAA|TTG
Hpa II	*Haemophilus parainfluenzae*	C|C*GC GGC*|C
Msp I	*Moraxella* species	C|CGG GGC|C
Hinc II	*Haemophilus influenzae* R_c	GTY|RAC CAR|YTG
Not I	*Nocardia otitidis-cavarium*	G|CGGCCGC CGCCGGC|G
Sfi I	*Streptomyces fimbriatus*	G|GCCNNNNNGGCC CCGGNNNNNCCG|G

* This sequence is recognised even if this C is methylated.

Several nucleases that will hydrolyse both DNA and RNA, provided they are single-stranded, are generally regarded as endonucleases but they are capable of degrading their substrates completely to 5′-nucleotides. They will digest away portions of DNA or RNA sequences that are not base-paired to a complementary strand. The most widely used of these enzymes is the S1 nuclease from *Aspergillus oryzae*.

Fig.3.6. Detection of start sites of transcription by S1 nuclease mapping.

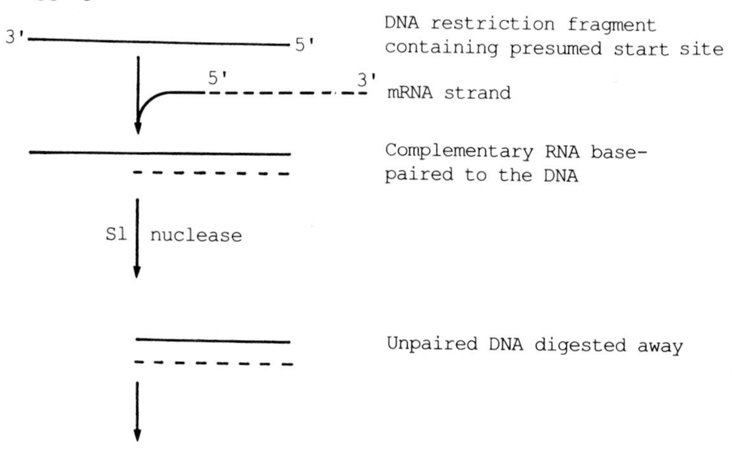

A very important use of S1 nuclease is in mapping the site at which transcription of DNA starts. A restriction fragment of DNA containing the presumed start site is prepared and labelled at its 5′-end (Chapter 3.13). This is hybridised (Chapter 3.12) to some of the mRNA whose synthesis it directs and treated with S1 nuclease to digest away any unpaired DNA. If the remaining DNA–RNA is now denatured, the DNA can be separated and sequenced to establish the 3′-end of the DNA at which transcription begins (Fig. 3.6).

DNA polymerases I and III both have exonucleolytic activity on DNA. Polymerase I excises damaged nucleotides (e.g. thymidine dimers formed by ultraviolet radiation) in the repair of DNA, prior to reformation of the correct sequence, while this activity of DNA polymerase III is important for proof-reading newly synthesised DNA (Chapter 5.1).

3.9 Restriction endonucleases

These are bacterial enzymes possessing an endonuclease activity directed to specific base sequences in double-stranded DNA. In nature, they protect bacteria from the possible incorporation of foreign DNA into their genomes by hydrolysing it. The bacterium's own DNA is protected by methylation on A or C residues which renders it resistant to attack by its own enzymes. The term 'restriction' is used because it was found that certain bacteriophages would not grow on certain bacterial strains – hence they were said to be restricted, due to the action of these enzymes.

Over 350 restriction enzymes are known, although only about 100 distinct sequences are recognised, since, in many cases, enzymes from different organisms recognise the same sequence. Such enzymes are known as *isoschizomers*. They are named by three letter abbreviations derived from the name of the species in which they occur, sometimes followed by a further letter or numeral to differentiate enzymes from the same species, e.g. Eco RI is an enzyme derived from *E. coli* strain R. The base specificity of several of these enzymes is shown in Table 3.3. Note that the sequences in the two strands of DNA at the recognition site possess a two-fold axis of symmetry, known as a *palindrome*. Some enzymes make cuts that are precisely opposite in the two DNA strands, so that the cut ends are said to be 'blunt': more usually the cuts are staggered, leaving a few unpaired bases at the end of each strand ('*sticky ends*'). This is very useful in joining two DNA fragments from different sources, since the complementary unpaired bases will stick together by hydrogen bonds (Fig. 3.7).

Hydrolysis of genomic DNA by any particular restriction endonuclease

Fig. 3.7. The association of Eco **RI** fragments from different DNAs by hydrogen bonding of their sticky ends.

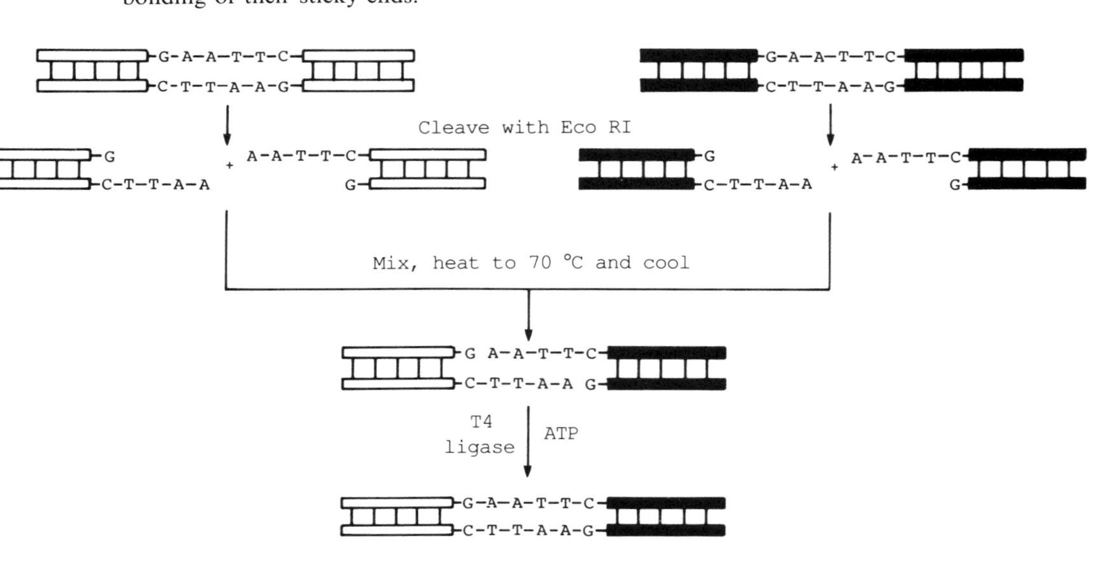

is likely to produce only a limited number of fragments since there will not be many bonds that can be attacked. Libraries of genomic DNA have been made by cloning into suitable vectors mixtures of fragments obtained from incomplete restriction enzyme digestion or by random shearing.

Complete digestion of DNA with a restriction enzyme may produce several smaller fragments whose lengths can be estimated by electrophoresis. The order of these fragments in the original DNA can be deduced by partial digestion with the same enzyme to produce some larger fragments. These are isolated and digested to completion to show which of the fragments in the complete digest they give rise to. Digestion of either the whole DNA or of these fragments with a second restriction endonuclease will produce further fragments that can also be ordered. In this way a *restriction map* can be built up. Figure 3.8 is a greatly simplified map of the plasmid pBR 322 on which there are 500 known restriction sites. Presence or absence of a restriction site can be predicted when the base sequence is known. Restriction maps are very useful in comparing portions of genomes since exact correspondences suggest that two sequences of DNA are identical or have arisen by very recent duplication of a portion of the genome. Loss or gain of a restriction site may result from a point mutation, a deletion or an insertion of nucleotides.

The position of overlaps in collections of genomic DNA can be established by restriction mapping (Fig. 3.9) so that it is possible to order

Fig. 3.8. A simplified map of the commonly used synthetic plasmid pBR322, showing a few of the sites of cleavage by restriction endonucleases, and the position of the replication origin (ORI), and the genes for ampicillin resistance (Ap) and tetracyclin resistance (Tc).

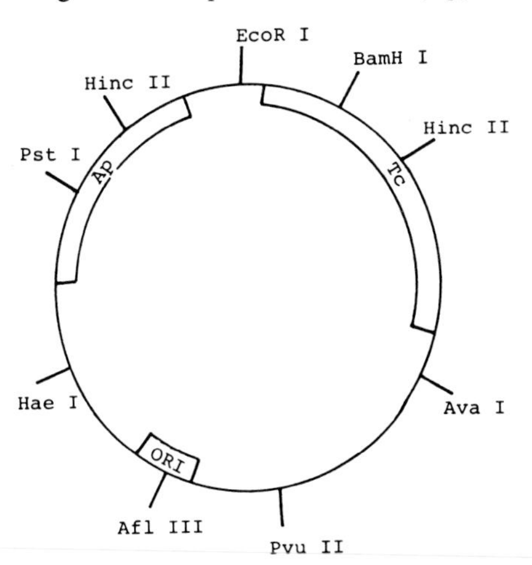

correctly much larger pieces of DNA than can be cloned in one piece. This approach to building up maps of relatively long stretches of DNA is known as '*chromosome walking*'.

3.10 Reporter genes

The effect of a potential regulatory sequence (Chapter 7.10) can be studied by placing under its control a 'reporter' gene, coding for an enzyme not normally expressed in the cell where it is being investigated. This gene is placed immediately downstream of the promoter of another strongly expressed gene. The effect on the expression of the reporter gene of deletions, insertions or mutations in the DNA upstream from its promoter can be readily assessed by estimating the amount of enzyme produced when the artificial construct is transcribed and translated in a suitable system.

In a typical construct a plasmid is engineered by insertion of a strong eukaryotic promoter such as that of the herpes simplex virus thymidine kinase gene immediately 5′ to the reporter gene. Genes that are often used for this purpose are those encoding chloramphenicol acetyl transferase and the β-galactosidase of *E. coli*. These both code for enzymes that are not normally found in eukaryotic cells, and can be readily detected by quantitative assays. Restriction sites 5′ to the promoter are used for inserting sequences derived from the regulatory sequences that are being

Fig. 3.9. Ordering of overlapping genomic fragments by restriction mapping. The horizontal lines represent cloned fragments of DNA from the region of the mouse genome encoding the fourth component of complement (Chapter 10.13). The vertical lines show the sites at which Bam HI acts. Other restriction enzymes were also used to confirm these results.

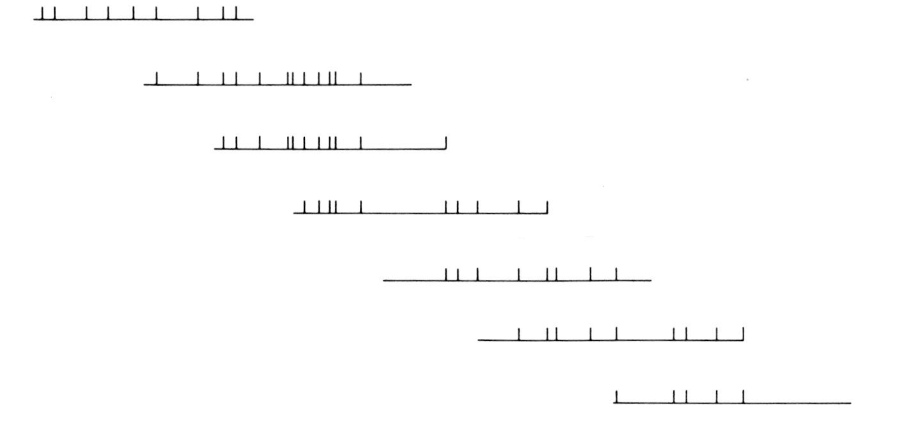

studied (Fig. 3.10). The plasmid is then incorporated by transfection into suitable cells (Chapter 4.1), grown under appropriate conditions (e.g. absence and presence of inducers) and the amounts of reporter enzyme produced are determined.

3.11 Restriction fragment length polymorphisms

Digestion of the whole human genome to completion with a restriction endonuclease produces a large number of fragments. These can be separated by electrophoresis and individual ones shown up by Southern blotting (Chapter 3.12), using a probe made from any desired part of the genome. For example, a probe from the genes of the β-globin cluster (Chapter 9.1) detects restriction fragments of 2·7, 3·5, 7·2 and 8·0 kbp after digestion with Hind III. DNA from many individuals suffering from sickle cell anaemia lacks the bands of 2·7 and 7·2 kbp because of mutations causing the loss of two sites where this enzyme acts. Since these sites are only present in the genome of certain individuals, they are polymorphic.

Polymorphisms detected in this way are known as Restriction Fragment Length Polymorphisms (RFLPs). They can arise from point mutations leading either to loss or gain of a site where a restriction endonuclease acts (Fig. 3.11). In addition, deletions or insertions can alter the length of sequences between two restriction sites. Restriction fragment length

Fig. 3.10. Map of a reporter gene incorporated into a plasmid. X: sequence being studied; TK-P Herpes simplex virus thymidine kinase promoter; CAT: chloramphenicol acetyl transferase gene; End: terminator sequence for eukaryotic virus; RE sites: the sites opened by a suitable restriction enzyme for insertion of X.

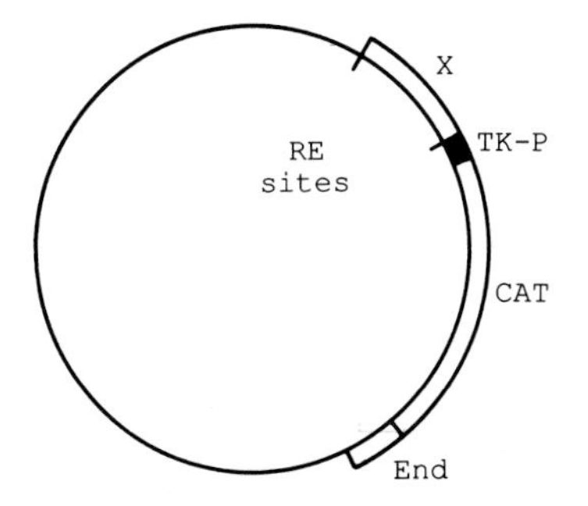

Fig. 3.11. Hypothetical point mutations leading to gain and loss of a Hind III restriction site (shown as the middle sequence).

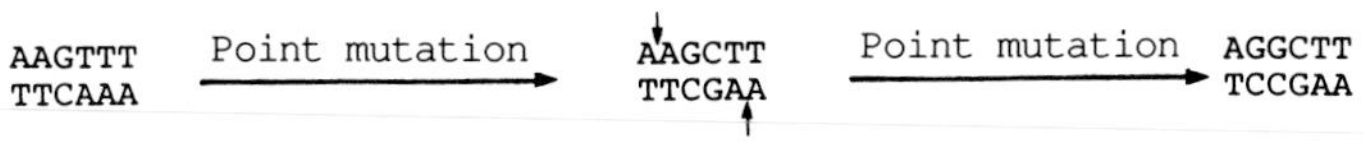

AAGTTT — Point mutation → AAGCTT — Point mutation → AGGCTT
TTCAAA — — — — — — — — — TTCGAA — — — — — — — — — TCCGAA

polymorphism is used for diagnosis of genetic disorders, especially when a foetus is at risk of being genetically disabled by inheriting defective genes from both parents. Note that these polymorphisms are not necessarily situated in that sequence of the genome coding for the harmful aberrant product (as in the example cited above). Provided they are closely and stably linked to such a mutant sequence they can be used to detect its presence. Foetal cells obtained by amniocentesis may have to be cultured for several weeks to produce enough DNA to make the test. A newer technique, which is still being evaluated for safety, involves removing some trophoblastic villi that will yield larger amounts of DNA sufficient for performing the test.

RFLPs have been reported that are linked to the inheritance of phenylketonuria and to Huntington's chorea, and doubtless it will be possible to detect many other inherited diseases by this technique in the future.

3.12 Hybridisation of nucleic acids

Double-stranded DNA can be denatured by various procedures that break the hydrogen bonds linking the complementary base pairs, such as heating it in aqueous solution. The dissociation of the two strands results in an increased absorbance at 260 nm, where nucleic acids have an absorbance peak. This provides a simple way of monitoring the extent of the separation, which occurs over a fairly small temperature range, frequently around 60 °C–80 °C, called the *melting temperature* (T^{o}_{m}). Since triple-bonded G–C pairs are more stable than A–T pairs, the greater the proportion of G–C in a sample, the higher its T^{o}_{m} is likely to be. Other agents that denature DNA are high salt concentration and various organic solvents such as formamide and dimethyl sulphoxide.

Denaturation is usually reversible so that, on cooling a heat-denatured specimen, the complementary strands reassociate. This process is known as renaturation or *annealing*, and can occur between DNA strands that were not originally base-paired, provided they are sufficiently complementary. RNA also hybridises to complementary single strands of DNA in an analogous way. Complexes between DNA and RNA are known as *heteroduplexes*, to distinguish them from *homoduplexes* formed from two complementary single strands of DNA. RNA–DNA heteroduplexes form preferentially in a mixture of denatured double-stranded DNA and RNA because they are more stable.

DNA absorbs strongly to nitrocellulose, so it can be spotted on to nitrocellulose paper and denatured and fixed by heating. Alternatively, prints can be made from gels after electrophoresis of DNA fragments.

Specific sequences in the DNA are detected by hybridisation to probes of radioactively labelled RNA or DNA at an elevated temperature in the presence of formamide. On cooling, the probe anneals to any DNA on the blot containing complementary sequences, and excess uncomplexed probe is washed away, or removed by digestion with S1 nuclease. The binding of the probe can be detected by autoradiography, or by direct counting of the filter. This technique is called *Southern blotting*, after the name of its originator. The complementary technique of *Northern blotting* hybridises single-stranded DNA to RNA which has been blotted on to nitrocellulose paper.

The precise conditions under which hybridisation is performed may affect the extent of hybridisation with similar but not identical sequences. *Stringent conditions*, using higher temperatures and higher concentrations of formamide only allow hybridisation of perfectly complementary sequences or, at best, those with very few mismatches. At lower stringency the formation of hybrids from sequences containing appreciable numbers of mismatching bases will occur. This is useful for detecting related, though not identical, sequences either in different parts of one genome or in homologous sections of the genome from different species.

DNA oligomers up to 50 nt long can easily be designed and synthesised to match the probable mRNA sequence coding for short portions of proteins whose amino acid sequences are known. These are used as probes in Southern blots to detect the genes for these proteins by a method known as 'reverse translation'. Because of the degeneracy of the genetic code, it will not be known which particular codon is used for each amino acid. This difficulty is overcome by choosing lengths of amino acids that have few alternative codons and by synthesising oligonucleotides into which appropriate bases are incorporated randomly at positions where the uncertainties exist. Figure 3.12 is a hypothetical example that illustrates this point.

Heteroduplexes form between mRNAs and the genomic sequences that

Fig. 3.12. Hypothetical synthetic probes that could be used for reverse translation. The oligonucleotide mixture is built up one nucleotide at a time except where degeneracy of the genetic code requires one of two or four bases to be randomly incorporated. Thus a mixture of $2 \times 2 \times 2 \times 4 = 32$ oligonucleotides is produced.

```
Met.Phe.Tyr.Glu.Trp.Pro.....

                       U
      U   U   A        C
ATG.UUC.UAC.GAG.UGG.CCA.....
                       G
```

encode them. Examination in the electron microscope shows the position of introns in the genes (Chapter 7.17). mRNA lacks these introns so they will be seen as bubbles of single-stranded DNA (Fig. 3.13), known as R loops. If size markers are included in the mixture spread for microscopy, the length of each intron can be measured to give an estimate of the number of bases it contains. Such measurements are usually accurate to about 50 bases.

3.13 The determination of base sequence in DNA

The determination of the base sequence in DNA is now a relatively easy procedure, thanks to the development of two methods, both of which produce a series of DNA molecules differing in length by a single nucleotide that can be separated by electrophoresis.

The method devised by Maxam and Gilbert uses chemical reactions to cleave the DNA at specific base residues (Fig. 3.14). Dimethyl sulphate methylates guanine residues so that they can be displaced by reaction with piperidine. This also eliminates the sugar residue to which the guanine was

Fig. 3.13. R loops formed by reaction of cloned DNA coding for mouse IgC (Chapter 10.4) with the corresponding mRNA. On the left is the actual electron micrograph. On the right is an interpretive drawing. A and B are double strands of DNA outside the region being studied. The dotted line is mRNA. V is the region coding for the variable part of the Ig chain. R1, R2 and R3 are loops formed when the mRNA anneals to one strand of DNA. I1 and I2 are introns where the two strands of DNA have re-associated . (Reprinted by permission from H. Sakano *et al.*, *Nature London* (1979), **277**, 627. © 1979 Macmillan Journals Limited.)

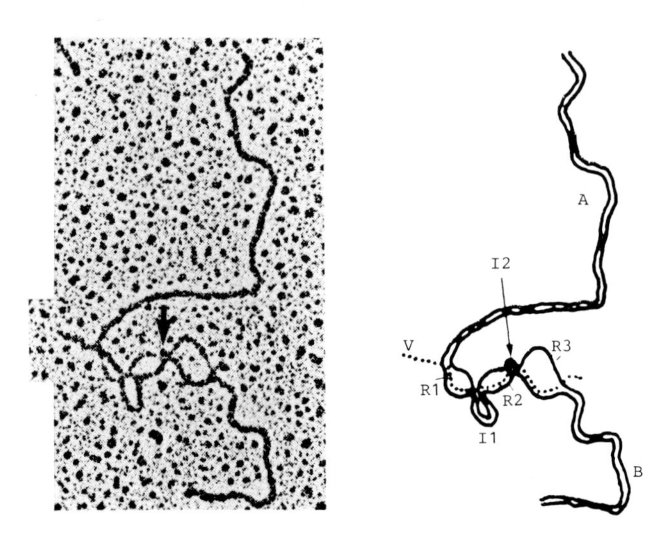

attached, and breaks the DNA chain at this point. Since the initial methylation reaction has an equal chance of occurring at any guanine residue, a series of fragments are produced, broken at each of these residues in the sample, provided the dimethyl sulphate is not in excess. Heating the DNA in acid solution specifically removes both purine bases and subsequent treatment with piperidine eliminates the deoxyribose, cleaving the chain at both adenine and guanine residues. Thymine and cytosine residues both react with hydrazine, and subsequently piperidine

Fig. 3.14. Selective cleavage of DNA at a guanosine residue by reaction with dimethyl sulphate and piperidine. Guanosine and adenosine residues are both removed by treatment with acid prior to reaction with piperidine, while thymidine and cytidine residues are lost from the DNA chain by reaction with hydrazine.

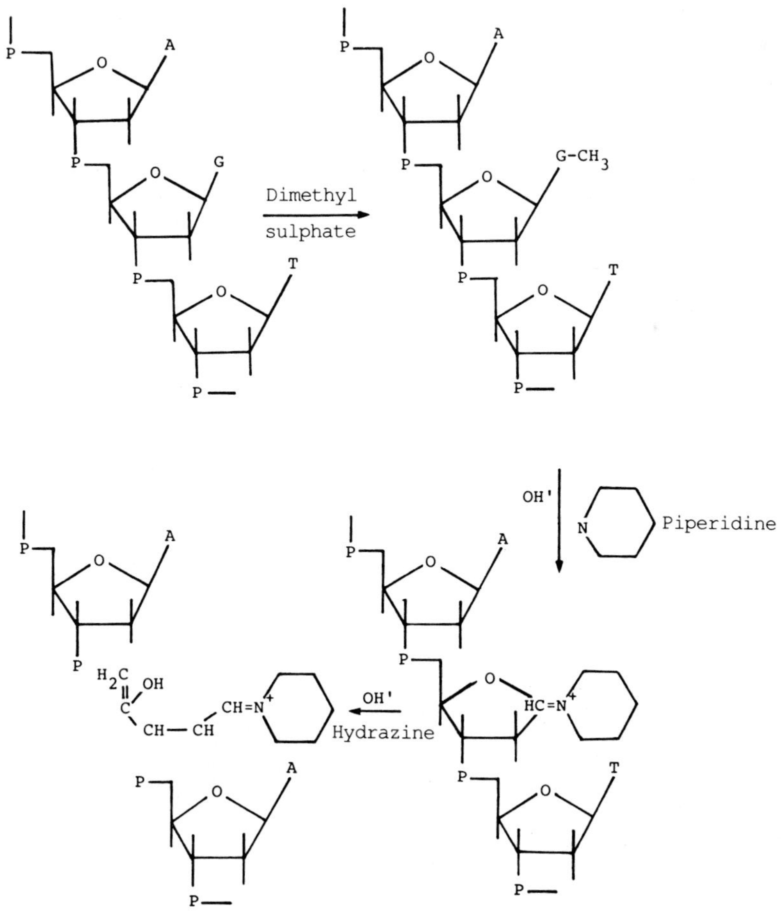

can break the chain at these two residues. Finally, in the presence of a high concentration of salt, hydrazine only reacts with cytosine residues, and subsequently piperidine cleaves at these residues only.

Thus, four reaction mixtures are set up simultaneously, and after reactions have gone to completion each mixture is electrophoresed. The fragments are separated according to their sizes, giving a series of bands (a '*ladder*') on each gel, and the sequence of bases can be read off by observing which treatment caused the breaks in the original DNA. In order to detect the oligonucleotides on the gel, the DNA is first made highly radioactive. The 5′-end can be labelled by treatment with alkaline phosphatase to remove the 5′-phosphate group, followed by incorporation of ^{32}P from γ-^{32}P-ATP catalysed by a polynucleotide kinase. Alternatively, the 3′-end is labelled by reaction with α-^{32}P-ATP, catalysed by terminal deoxynucleotidyl transferase (Fig. 3.15). Oligonucleotides up to about 250–300 nt long can be separated by electrophoresis. In practice, it is usual to run two gels – one for a short time to separate the smaller oligonucleotides, and one for longer to separate the larger ones. Figure 3.16 shows a typical sequence ladder. Although the method is limited to about 300 nt, very much longer sequences can be determined by first cleaving the DNA with restriction endonucleases to give overlapping fragments, and then sequencing each one.

A second method of sequence determination, developed by Sanger and his colleagues, involves the enzymic synthesis of stretches of DNA on a

Fig. 3.15. End labelling of DNA: (a) at 5′-end, using alkaline phosphatase, polynucleotide kinase and ATP, labelled on the γ-P atom; (b) at 3′-end, using terminal deoxyribonucleotidyl transferase and ATP, labelled on the α-P atom. P in this figure symbolises the phosphate group.

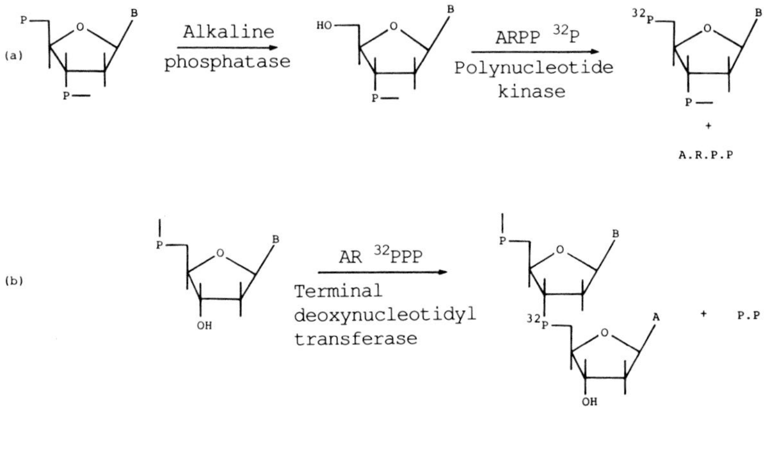

template of the DNA whose sequence is sought. Newly synthesised fragments are terminated at random by a nucleotide analogue that blocks further additions of nucleotide residues. This is usually a dideoxynucleoside triphosphate (Fig. 3.17) which cannot accept the next nucleotide in the sequence because it lacks a 3′-OH group.

A series of genetically engineered derivatives of the bacteriophage M13

Fig. 3.16. Sequence ladders used in DNA sequencing by the technique of Maxam & Gilbert. Left: ladder for determination of sequence of part of the vasopressin-neurophysin II gene in normal rats. Right: ladder for determination of sequence for part of the vasopressin-neurophysin II gene in rats of the Brattleboro strain with diabetes insipidus (D.I.) (Chapter 11.7). The deduced sequences are shown at the bottom. The arrow in the right-hand ladder shows where the deleted G is missing. (Reprinted by permission from H. Schmale & D. Richter, *Nature London* (1984), **308**, 706. © 1984. Macmillan Journals Limited.)

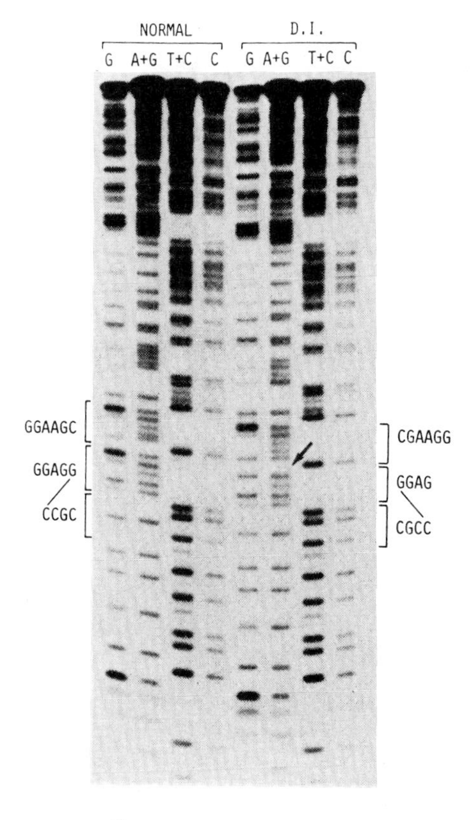

(Chapter 4.2), which contains a single-stranded circular genome, have been constructed with contiguous sites for a number of restriction endonucleases (Fig. 3.18). This DNA can be opened at one of these sites and ligated to the DNA whose sequence is to be determined. Next, a synthetic oligonucleotide primer with a sequence complementary to about 15 bases on the 3′-side of the restriction sites is annealed to the phage. This is added to four parallel reaction mixtures, each containing three of the deoxyribonucleoside triphosphates, plus the fourth (usually dATP) radioactively labelled, and the Klenow fragment of DNA polymerase I. This fragment lacks the exonucleolytic activity of the enzyme, but retains the polymerase activity and catalyses the synthesis of a new DNA strand complementary to that which has been inserted into the M13 vector. Four sets of fragments are produced, with all those in any one set terminating at a particular residue. They are separated as a ladder by electrophoresis, just as in the Maxam and Gilbert method.

Fig. 3.17. Sanger's dideoxy chain terminating method for sequence determination.

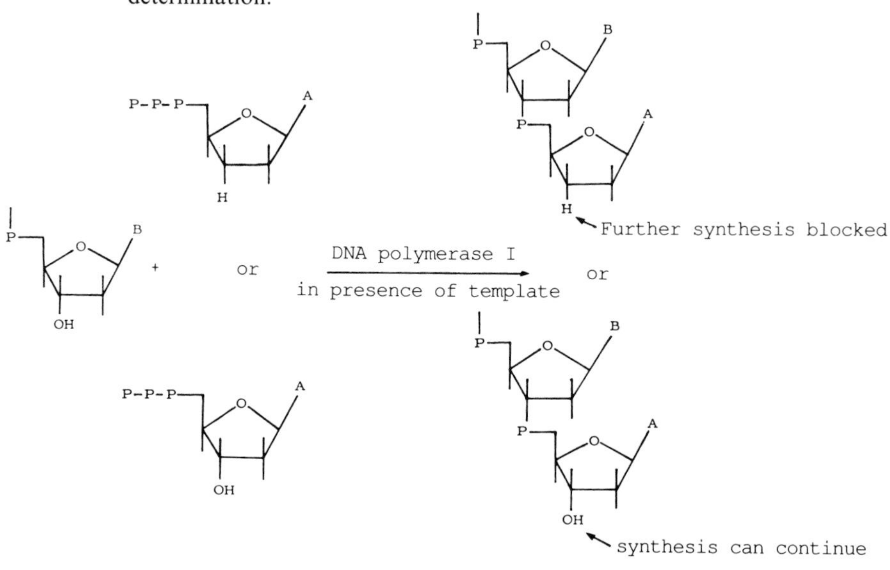

Fig. 3.18. Sequence of nucleotides introduced into the phage M13mp8, which is often used in Sanger's sequencing method. Several restriction sites are available, and a 17-nucleotide stretch has a complementary oligonucleotide which is available as a primer.

Sma I — Hinc II — Primer

—GAATTCCCGGGGATCCGTCGACCTGCAGCCAAGCTTGGCACTGGCCGTCGTTTTAC—

EcoR I — BamH I — Pst I — Hind III

The key to the success of both methods lies in the fact that the sites of specific cleavage or chain termination are completely random. This is achieved by ensuring that the chain-breaking or chain-terminating reagents are not in excess.

Figure 3.19 shows the fragments that would be generated from a hypothetical short length of DNA by both methods.

The ease and simplicity of these elegant methods means that it is generally easier to sequence DNA than protein. The sequences of many proteins have been inferred from the sequences of either their genes or the cDNAs made from their mRNAs.

In an alternative to the Sanger sequencing method the primer DNA is labelled at the 5′-end by chemical coupling with a fluorescent dye, rather than a radioactive isotope. Four fluorescent compounds with different excitation and emission wavelengths are coupled to separate samples of the primer. Then four sequencing reactions with dideoxynucleotides are performed as usual, each using one of the fluorescently labelled primers, and the products are electrophoresed in a narrow tube of gel. The eluate from this is scanned by a fluorimeter that operates in rapid and cyclic

Fig. 3.19. Determination of the sequence of a simple oligonucleotide by (a) Sanger's method, and (b) Maxam & Gilbert's method, showing the products that will be obtained by the two procedures.

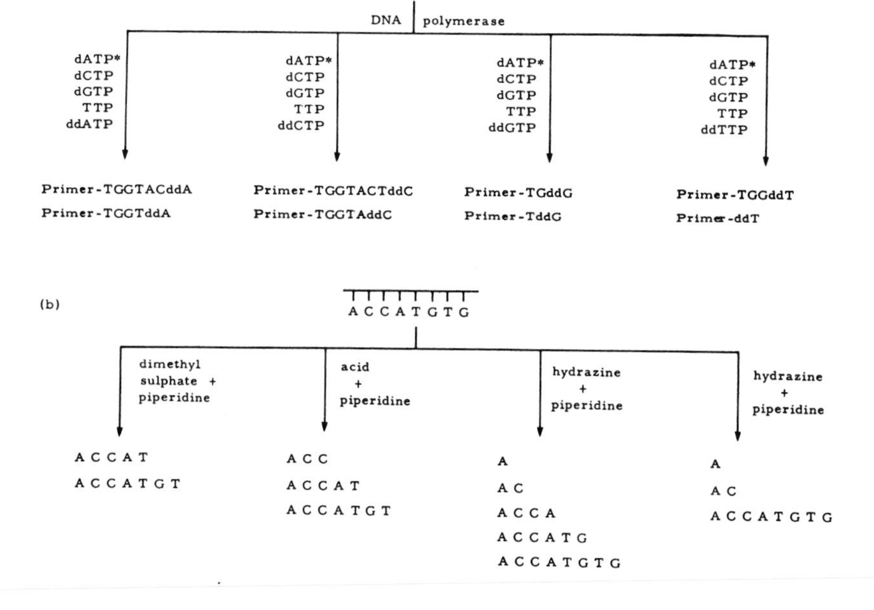

succession at four appropriate excitation and emission wavelengths. This produces a record of which particular extended primer is eluted at any one point and hence which nucleotide occurs at that position.

A great advantage of this method is that it is adaptable to automation, but at present it is only about an order of magnitude less sensitive than the original method using radioactive markers, so that larger amounts of DNA are required. However, it is likely that the sensitivity can be increased.

4

Vectors used in work with recombinant DNA

DNA fragments or cDNAs that have been produced as described in the last chapter can be propagated by cloning them in suitable hosts. This involves ligating the DNA covalently to a suitable vector, which can then be manipulated so as to multiply in an appropriate type of host cell. Vectors include plasmids and bacteriophages, both of which will grow and replicate in bacteria, and viruses which can be grown in eukaryotic cells.

4.1 Plasmids

Plasmids consist of double-stranded DNA, found in bacteria, but generally not associated with the bacterial chromosome. They are stably inherited and utilise a specific region of their genome (the replication origin) for autonomous replication. Under natural conditions they will usually replicate to give 20–30 copies per cell, but if, for example, the bacteria are grown on a medium containing chloramphenicol yields of up to 1000 or more copies per cell can be obtained. They generally exist as covalently closed circular DNA molecules, smaller than the bacterial chromosome by 2–3 orders of magnitude (i.e. they contain 10^3–10^5 base pairs). Under suitable conditions their DNA can be broken open to yield a linear molecule which passes from one bacterium to another in a process known as transfection. They can be isolated in a relatively pure state from a bacterial lysate by centrifuging in a density gradient.

The DNA of plasmids is transcribed to give mRNAs which direct the synthesis of corresponding proteins. Some plasmids (R factors) carry genes for antibiotic resistance, specifying an enzyme that will detoxify the antibiotic by various chemical reactions, or interfere with its uptake (Table 4.1). Other types of plasmid carry genes for the synthesis of proteins known as bacteriocins that are lethal to other bacteria not

Table 4.1. *Mechanisms by which plasmids confer resistance to some antibiotics*

Antibiotic	Mechanism
Chloramphenicol	acetylation by chloramphenicol-acetyl transferase (CAT)
Tetracyclin	interferes with uptake
Ampicillin	hydrolysis by β-lactamase
Kanamycin	induction of transferase

Table 4.2. *Some properties of commonly used plasmids*

Plasmid	Size (bp)	Genetic markers*	Restriction sites†
pBR 322	4362	Ap^R, Tc^R	EcoRI Hind III (Tc) Pst I (Ap)
ColE1	4200	colicin production	EcoRI (CP) Sma I (CP)
pCR1	7400	colicin immunity, Km^R	EcoRI Hind III (Km)
pACYC 184	4000	Cm^R, Tc^R	EcoRI (Cm) BamHI (Tc)
pJC 74‡	16000	Ap^R	EcoRI Bgl II Pst (Ap)
pJC 79‡	6000	Ap^R, Tc^R	EcoRI Pst I (Ap) Hind III (Tc)

* Ap^R = ampicillin resistant; Tc^R = tetracyclin resistant; Km^R = kanamycin resistant; Cm^R = chloramphenicol resistant.
† When the restriction enzyme is followed by a symbol for an antibiotic, this indicates that it cuts within the gene specifying resistance to that antibiotic.
‡ Cosmid (see Chapter 4.2).

carrying the same type of plasmid. The best known ones are the colicins that are made by plasmids such as ColE1 in *E. coli.*

Many naturally occurring plasmids have been isolated and modified in various ways. Portions of their DNA can be deleted to make them smaller; genes can be introduced to provide products to serve as markers; regulatory elements (Chapter 7.10) that may respond to the presence of certain metabolites in the medium can also be inserted. It is obviously essential that the region around the replication origin should not be altered by these procedures. Some details of a few plasmids are shown in Table 4.2.

Plasmids can be cleaved by restriction endonucleases. Figure 3.8 shows a greatly simplified restriction map of the plasmid pBR 322 on which there are 500 known restriction sites. Even with this large number of sites, there are at least 20 restriction endonucleases that will not cleave this plasmid.

Recombinant DNA can be prepared by cleavage at one of these restriction sites. If this leaves short lengths of unpaired bases – 'sticky ends' – and if a fragment of DNA derived from some other source (*passenger* DNA) by treatment with the same enzyme is mixed with the cleaved plasmid, some of these sticky ends of the plasmid will associate by hydrogen bonding with the restriction fragment. The passenger DNA and the plasmid are joined covalently by DNA ligase from either *E. coli* or the bacteriophage T4 to form a new recombinant DNA molecule (Fig. 4.1). An alternative method is to cleave both plasmid DNA and the passenger DNA to be cloned with a restriction endonuclease (e.g. Hae III) which yields 'blunt-ended' cuts. These can also be joined together by the action of the T4 ligase, though not by the enzyme from *E. coli*. In both these methods undesired ligations (i.e. passenger to passenger or vector to vector) may be introduced.

Blunt-end ligations are also used to introduce sites at which specified restriction endonucleases can act at a later time. This is achieved by ligating synthetic oligonucleotides, known as linkers, to DNA fragments (Fig. 4.2). Some of these linkers are fairly complex, and may contain sites for several restriction endonucleases.

Fig. 4.1. Construction of a cloned vector in the plasmid pBR322. The passenger DNA would have been isolated after cleavage of a large DNA by the restriction endonuclease BamH I. Ap^R and Tc^R are the sites of the genes for ampicillin resistance and tetracyclin resistance respectively. The latter gene will have been inactivated by the cloning of the passenger DNA.

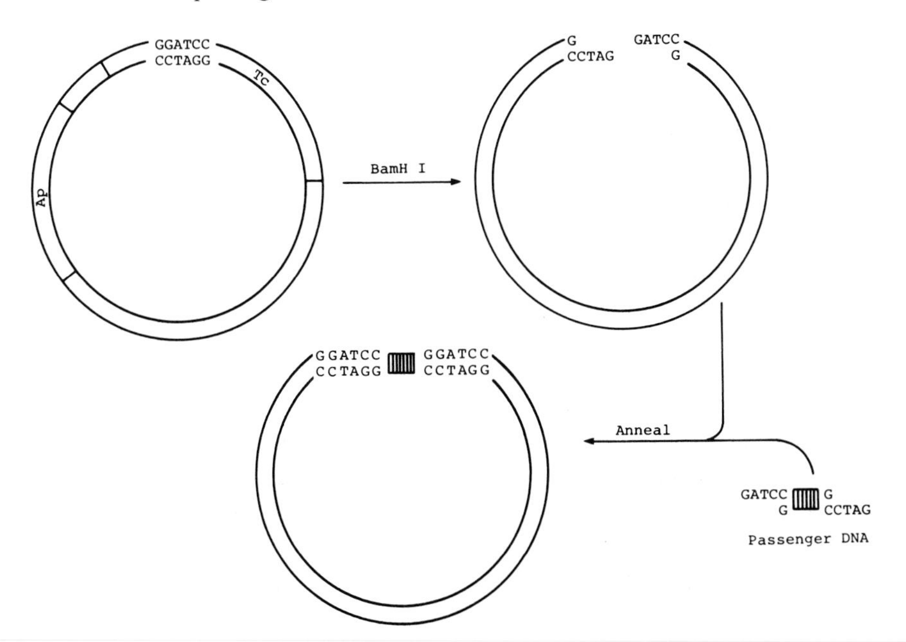

Two blunt-ended pieces of DNA can also be linked together by building up homopolymers on their 3′-ends with the enzyme terminal deoxynucleotidyl transferase (Fig. 4.3.). Reaction conditions are chosen so that only a small number of nucleotides are incorporated. The added bases are a complementary pair (e.g. G to the plasmid DNA and C to the passenger DNA). The two DNAs are annealed, and the hybrid that is formed is stable enough to be used directly for transfection. Any gaps are filled in by the host's own synthetic enzymes. Since there will be no circularisation of cleaved plasmids with homopolymer tails, only the hybrid molecules are infectious, so all bacteria containing the plasmid also contain the inserted DNA. However, the infectivity of these annealed forms is generally lower than that of covalently closed forms.

The cloning of plasmids containing passenger DNA into bacteria is known as *transfection*. It is usually achieved by mixing suspensions of the

Fig. 4.2. Introduction of restriction sites into DNA by the use of synthetic linker oligonucleotides.

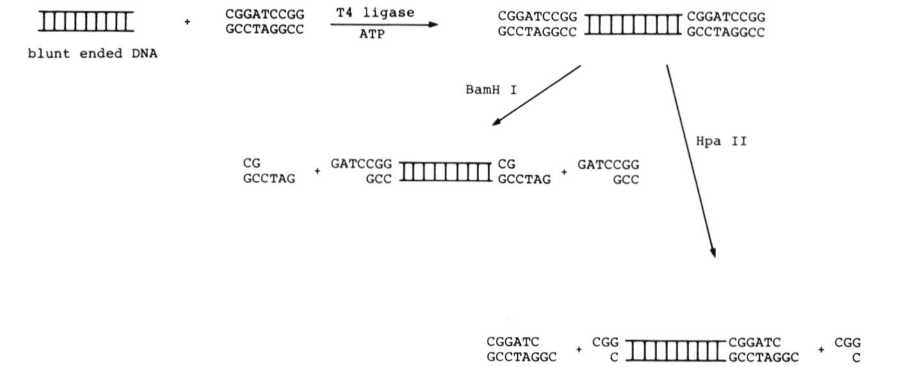

Fig. 4.3. Incorporation of passenger DNA into a vector using homopolymer addition to sites generated by cleavage with a restriction endonuclease.

— CTGCAG —
— GACGTC —
vector
Pst I
— CTGCA G —
— G ACGTC —
Terminal deoxyribonucleotidyl transferase
nGTP
— CTGCA(G)n G —
— G (G)nACGTC —
Genomic DNA fragmewnt
Terminal deoxyribonucleotidyl transferase
nCTP
(C)n
(C)n
anneal
— CTGCA(G)n (C)n G —
— G (C)n (G)nACGTC —
Filling in and Ligating with host enzymes
— CTGCA(G)n (C)nTGCAG —
— GACGT(C)n (G)nACGTC —

plasmid and the bacterial host in the presence of a relatively high concentration of calcium (20–50 mM). Since not all the plasmids will contain the passenger DNA, the bacterial culture must be screened for those that do. Firstly, the bacteria are grown in a medium containing an antibiotic against which the plasmid confers resistance. This will kill all the bacteria that have not been transfected. Many of the survivors containing the plasmid will not contain the passenger DNA. If this has been inserted at a site in a second gene conferring resistance to a different antibiotic, this property will have been lost, and can be used to select the cells that do contain the passenger DNA.

Other methods may be used to select cells that actually contain the desired DNA sequences. For example, if the bacteria are spread on a nitrocellulose filter supported on a nutrient agar plate and allowed to grow, a replicate set of colonies will be produced. The filter is then removed and washed, and the cells on it are lysed by a detergent to release the DNA which is fixed to the nitrocellulose by drying at 80 °C. The filter is then soaked in a solution of 50% formamide and an isotopically labelled RNA probe which is known to hybridise with the desired DNA sequence. After washing and drying, autoradiography will reveal any clones that contain this sequence. These colonies can be picked off the replicate plate and grown on to provide any desired quantity of the DNA.

Finally, the DNA is recovered from the amplified plasmids by lysing the bacteria, usually with lysozyme followed by a detergent. The plasmid DNA is separated from the chromosomal DNA by centrifuging on a density gradient of caesium chloride on which the plasmid DNA sediments faster than the bacterial DNA. Digestion of the plasmid DNA with a suitable restriction endonuclease generates the desired fragments that can be separated from other DNA fragments by electrophoresis.

There is also a yeast plasmid, referred to as the 2 μm plasmid (6·3 kbp long). It can be manipulated in similar ways to bacterial plasmids to provide vectors that will grow and replicate in yeast cells. Some hybrids of bacterial and yeast plasmids are actually capable of growing in either organism (shuttle vectors), and these are valuable tools for the study of gene expression.

4.2 Bacteriophages

Alternative vectors for amplifying selected DNA sequences are derived from bacteriophages (frequently referred to as phages). These are viruses whose specific hosts are bacteria. Their genome of DNA (occasionally RNA) encodes a number of proteins. Phages will only replicate in a bacterium of the correct species or strain. They may eventually kill an infected cell when a sufficient number of new phage particles have been synthesised. The cell will lyse so that the phages are released and can infect other cells. Temperate phages can infect a cell and replicate and lysogenise the cell without causing lysis. Virulent phages always replicate and lyse the host cell.

Phages have some advantages over plasmids as vectors for producing recombinant DNA:

1. it is comparatively easy to screen very large numbers of phage particles for a given DNA sequence by nucleic acid hybridisation;
2. they can infect bacteria more efficiently than plasmids;
3. large numbers of DNA fragments derived from a single large genome can be packaged, replicated and stored as a 'library' containing very many DNA sequences;
4. on the whole, larger DNA fragments can be cloned into phages than into plasmids.

Two phages and their derivatives have been widely used as vectors. Phage M13 is a virulent phage containing its DNA as a single-stranded circular molecule. Some derivatives are particularly useful in Sanger's method for determining the base sequence of DNA (Chapter 3.13).

Phage λ (lambda) is a temperate phage with a double-stranded linear DNA molecule. This is slightly less than 50 kbp long (just over 1% of the length of the *E. coli* chromosome), and many derivatives are available. In the lysogenic mode, its DNA integrates into a specific site on that chromosome. About a third of the phage DNA can be removed without impairing its ability to replicate and may be replaced by foreign DNA before infection of a host with the modified phage, provided there are unique restriction sites in the dispensable region of the genome. Wild type λ has five cleavage sites for Eco RI, and mutants have been selected that have lost three or four of these, leaving only one or two in the dispensable part of the genome. Passenger DNA can be inserted after cleavage at these sites in the same manner as into plasmids. Suitable bacteria are then infected with the altered phage DNA. Naked DNA will enter bacterial cells by transfection in the presence of a relatively high concentration of Ca^{2+}, but the uptake of phage DNA is very inefficient. This can be greatly improved by packaging the DNA together with the protein components of the phage head so as to make a complete bacteriophage. This is usually done by mixing it with two strains of phage, each lacking one of the head proteins, so that they complement each other. The complete phage now assembles itself and is used to infect a bacterial culture. The DNA can only be packaged if its length is between 75% and 109% of the wild type phage genome. In practice, the maximum size of passenger DNA that can be incorporated is about 23 kbp. This is above the size of many genes, but because of the large amount of non-coding DNA in most genomes it is unlikely to encompass more than a few complete genes.

After infection, the DNA is circularised by the action of the host's own DNA ligase. This is a necessary step before the phage DNA becomes integrated into the bacterial chromosome (Fig. 4.4). In this form the phage DNA replicates in the lysogenic mode at the same time as the bacterial

Fig. 4.4. Integration and excision of λ from *E. coli* chromosome.
c and c′ are cos sites: 5′-AGGTCGCCGCCC-3′,
3′-TCCAGCGGCGGG-5′.
b′, b and p′, p are homologous sequences:
5′-GCTTTTTTATACTAA-3′,
3′-CGAAAAAATATGATT-5′.

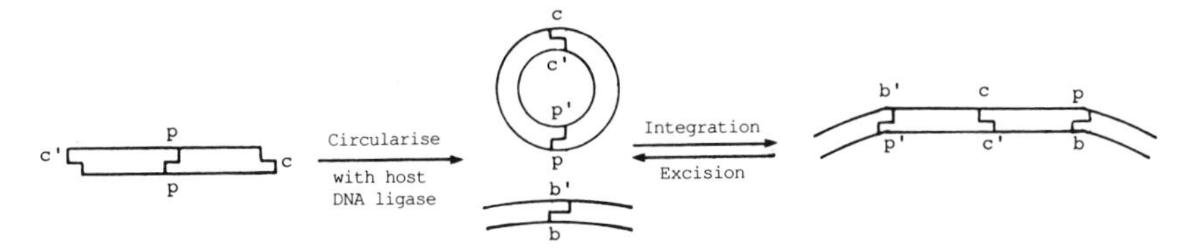

Fig. 4.5. Construction of recombinant DNA with a cosmid. ● = *cos* site. Note that concatamers of many different structures will be formed. Those of type A can be acted upon by λ terminase, and will be packaged efficiently. Those of types B, C and D (and others not shown) will not. On B the *cos* sites are too close together, while C lacks them. (Modified from J. Collins *Methods in Enzymol.* (1979), **68**, 310, Fig. 1; reprinted by permission of John Wiley & Sons, Inc., © 1979.)

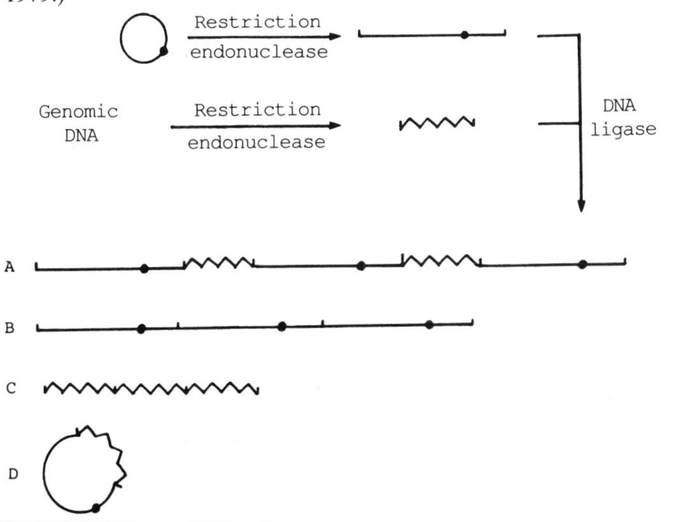

DNA. Under suitable conditions the phage can be induced into the lytic phase, when it is excised from the bacterial chromosome, and eventually lyses the cell, releasing many new phage particles.

At each end of the double-stranded linear DNA of λ is a short unpaired sequence of 12 nucleotides. Since these are complementary to each other (Fig. 4.4) they can form base pairs, and are known as cohesive ends (*cos*). When λ enters the lytic phase during its natural life cycle, these *cos* sites pair with each other and build up long chains (concatamers) of DNA. During the packaging of the DNA into the phage heads these are hydrolysed by an endonuclease called λ-terminase that recognises *cos* sites that are 38–52 kbp apart, producing DNA chains of this size. The packaging system described above only requires that the DNA contains the *cos* sites at a suitable distance apart: it is indifferent to the rest of the DNA sequence. Advantage has been taken of this to construct vectors called *cosmids* that are specialised plasmids into which *cos* sites have been engineered. A series of cosmids is available, each containing different restriction sites and a variety of antibiotic resistance genes that can be used for selection of the bacteria containing the desired cosmid. For use as vectors they are first cleaved by a restriction endonuclease and ligated to passenger DNA. During this process conditions are chosen so that formation of concatamers is favoured and occurs in high yield (Fig. 4.5). Because of the size limitation for packaging DNA into a λ head, these concatamers need to have *cos* sites about 38–52 kbp apart. Since a minimum of about 4–6 kbp of plasmid DNA is required for the construction of cosmids, inserts of 30–45 kbp can be accepted.

After packaging and infection of a suitable host the cosmid behaves as an autonomously replicating plasmid. The advantages of using cosmids instead of ordinary plasmids are:

1. virtually all the DNA that is packaged contains the passenger DNA so that screening of the progeny for hybrid clones is unnecessary;
2. there tends to be selection for large passenger DNA, whereas ordinary plasmids show a bias towards smaller fragments.

Since cosmids are smaller than the λ vectors this system has the potential for packaging larger amounts of passenger DNA than the latter.

4.3 Yeast artificial chromosomes

A serious limitation of the vectors derived from both plasmids and phages is that they will only accommodate passenger DNA up to about 50 kbp in length. Since many functional genetic units are a good deal longer, there is a need for vectors into which larger pieces of DNA can be incorporated. The recent introduction of yeast artificial chromosomes has made this possible. These constructs require DNA sequences containing *centromere* and *telomere* structures and *autonomously replicating sequences* (ARSs) so that the artificial chromosomes can replicate and multiply when introduced into yeast cells. The centromere is a DNA segment to which subcellular fibres attach during mitosis so that the replicated chromosomes separate properly. The telomeres are sequences at each end of a chromosome that apparently protect adjacent sequences from nucleolytic degradation. An ARS (Chapter 7.4) is required as an origin of replication. Artificial chromosomes have been constructed as hybrids between sequences derived from a plasmid such as pBR 322 and sequences from yeast. These are circular pieces of DNA, containing a centromere sequence, an immediately adjacent ARS and the TRP1 gene, as well as two telomeric sequences, a second yeast gene such as URA3 and a third yeast gene coding for an easily recognisable phenotypic marker with a restriction site into which the passenger DNA can be inserted (Fig. 4.6). Copies of the artificial chromosome are introduced into yeast cells

Fig. 4.6. Schematic map of yeast artificial chromosome. C = centromere; T = TRP1; U = URA3; A = Autonomous Replication Sequence. The single line represents sequences derived from a bacterial plasmid, including the origin and a gene for antibiotic resistance. The radial lines pointing outwards are sites for restriction endonucleases. The top one is used for inserting passenger DNA, and the bottom two are used to linearise the chromosome at the distal ends of the telomeres.

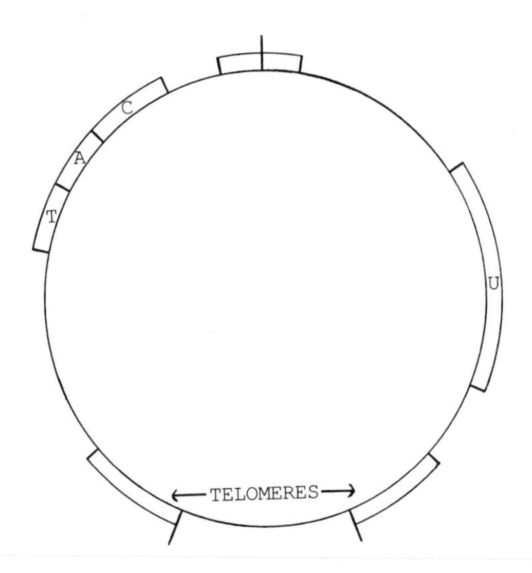

that are unable to grow on media deficient in tryptophan and uracil because they lack functional copies of the TRP1 and URA3 genes. The presence of these genes allows those cells that have taken up the artificial chromosome to grow on such media. The insertion of passenger DNA into the cloning site on the artificial chromosome is signalled by loss of the phenotypic marker. The circular DNA must be linearised before incorporation into yeast cells, and this is achieved by the action of a restriction endonuclease whose recognition sequences have been inserted at the distal end of each telomere.

The passenger DNA should be at least 50 kbp long, and stretches of up to several hundred kbp are easily accepted. In fact, artificial chromosomes seem to replicate and segregate better the longer they are.

In addition to being able to clone larger DNA sequences than would otherwise be possible, cloning with artificial chromosomes allows some post-translational modifications such as glycosylation to occur when the genes are expressed in yeast. This does not happen when eukaryotic genes or cDNAs are transcribed in prokaryotic systems.

4.4 Viruses

Viruses specific for eukaryotic cells can also be used as cloning vectors. The Simian virus 40 (SV40), a member of the group of papovaviruses, has been used in this way. It has a small genome of about 5 kbp. Half of this is transcribed early in its life cycle, and half late: the late gene region codes for viral structural proteins and is not all essential for viral replication. Portions of this can be cut out by restriction endonucleases and passenger DNA inserted. Cultured monkey cells can be either transfected directly with the recombinant DNA or infected with reconstructed viral particles containing the viral proteins as well as the recombinant DNA.

SV40, either as such, or carrying passenger DNA, can transform some rodent cell lines by a process in which the DNA is integrated into the host chromosome in a stable fashion. Other viruses such as the polyoma viruses, which are related to SV40, and the adenoviruses can also be used as vectors in a variety of appropriate mammalian host cells.

Experiments using these eukaryotic vectors are likely to be of more value than those using plasmids propagated in bacteria in studying the factors that are required for expression of genes in eukaryotic cells.

Baculoviruses, that infect the larvae of certain insects, are also used as vectors for expressing the products of eukaryotic genes. These viruses have double-stranded DNA genomes of 80–200 kbp, and it is believed that up to 100 kbp of foreign DNA can be inserted into and expressed from them. This passenger DNA is placed in a plasmid under the control of a viral promoter that is usually used for the transcription of an abundant viral protein. When suitable insect cells are transfected with a mixture of such a plasmid and some native viral DNA, recombination between these two DNAs ensures that some sequences of the plasmid DNA are replaced by viral DNA. Those DNAs that now contain the passenger DNA are selected by hybridisation to a suitable probe and then transfected into insect cells in culture. When these are grown up, the desired protein is synthesised and released.

Alternatively, the recombinant virus can be fed to insect larvae as this is the way in which the virus is naturally transmitted. In this situation the virus multiplies and eventually kills the larvae, releasing large numbers of new viral particles and also considerable amounts of the desired protein.

A commonly used insect is *Bombyx mori*, the silkworm, which can be cultivated in large numbers so there is the potentiality for producing virtually unlimited amounts of any protein whose gene or cDNA is available.

4.5 The production of genetically engineered proteins

If cDNAs to eukaryotic genes are introduced into bacteria those that are infected will produce the proteins that are encoded by the cDNAs. This technique has been applied to the production of human insulin by suitably engineered vectors in bacteria. However, since insulin consists of two polypeptide chains that are synthesised as a single chain precursor, proinsulin must be isolated and correctly cleaved to provide the mature and active hormone. This process makes virtually unlimited quantities of insulin available for therapeutic use.

Eukaryotic proteins produced in this way by bacteria will not undergo normal post-translational modifications such as glycosylation, and so may not possess their usual biological activity. It is often possible to overcome this limitation either by using vectors that proliferate in yeast or by the use of engineered vectors in cell cultures of higher animals.

5

Prokaryotic gene organisation and expression

Prokaryotes have a single chromosome consisting of circular double-stranded DNA. The size may vary considerably, e.g. the *E. coli* chromosome contains about $3{\cdot}8 \times 10^6$ bp, while that of *Bacillus subtilis* is about half that size (2×10^6 bp), and that of *Salmonella typhimurium* $10{\cdot}5 \times 10^6$ bp. Phages and plasmids have chromosomes which are up to three orders of magnitude smaller than these. If the *E. coli* chromosome were in a linear extended form it would be about 1 mm long. In fact the DNA is a fairly compact molecule due to supercoiling (Chapter 1). This means that it must be opened up to allow access for enzymes involved in replication and transcription. This is a complex process requiring several proteins including a Class I topoisomerase (Chapter 1.4) that relaxes supercoiled DNA, and single strand binding protein(s) (SSBP). The latter bind to single-stranded lengths of DNA produced by the action of the topoisomerase to keep them in open structures that are more accessible to the enzymes catalysing reactions on a single strand. Other proteins involved in these processes have been identified by genetic means but in some cases their actual roles are not known. They are named after the gene that codes for them (e.g. *dnaB*, *dnaC*, etc).

5.1 Replication

DNA replication begins at a definite site on the *E. coli* chromosome called oriC. A stretch of just over 1000 nucleotides in this region has been sequenced, and shorter sequences round the corresponding site in several other prokaryotic chromosomes are also known. A minimum of 245 bp is required to function effectively as a replication origin when this portion of the genome is engineered into plasmids. However, a longer sequence of at least 440 bp is necessary for bi-directional replication of both strands of the DNA. A consensus sequence can be derived from the known sequences of the different species in which short completely conserved stretches of up to 12 bp alternate irregularly with slightly longer unconserved sequences of up to 16 bp. These latter may serve as spacers between the binding sites of the various proteins involved in the initiation of replication. Several of the conserved sequences are inverted repeats which could form stem structures which might be involved in interaction with proteins.

In the OriC region there are four highly conserved nonamers TTAT$^{C}_{A}$CACA to which the dnaA protein binds in an ATP-dependent manner to provide an 'open' complex for binding the dnaB and dnaC proteins. dnaB protein is a helicase that starts unwinding the DNA duplex. This dissolution of the duplex is stabilised by SSBP that binds non-specifically to a sequence of about 33–65 bases in the single-stranded DNA in a co-operative manner. This generates a small bubble where the various other proteins required for replication bind.

DNA synthesis is a processive process, that is to say, once it is initiated it proceeds steadily along the chromosome from oriC. The actual site at which synthesis is occurring, where the strands are separated, is known as the *replication fork*. Since synthesis occurs in both directions simultaneously, there will be two replication forks at any one time.

As the replication fork opens up, one strand is in the correct ($3' \rightarrow 5'$) orientation for the continuous synthesis of the complementary strand in the $5' \rightarrow 3'$ direction catalysed by the DNA polymerase. This is known as the *leading strand*. The other ($5' \rightarrow 3'$) strand, known as the *lagging strand*, is wrongly oriented to provide a template for the continuous action of the polymerase. On this strand, DNA synthesis still takes place in the $5' \rightarrow 3'$ direction, but it is discontinuous, occurring in lengths of about 1000 nt at a time. The fragments of DNA produced in this way have actually been detected, and are known as Okazaki fragments, after their discoverer.

DNA replication must be initiated ('primed') with a short length of RNA that must be synthesised first. RNA polymerase probably provides the *primer* for the leading strand. Two promoters at which RNA transcription could be initiated are present in the oriC region, running in opposite directions. Also the initiation of replication is inhibited by rifampicin, an antibiotic inhibitor of RNA polymerase. However, the synthesis of primer on the lagging strand is probably performed by DNA primase (the product of the *dnaG* gene) that functions as a DNA-dependent RNA polymerase.

The synthesis of new strands of DNA is catalysed by a DNA-dependent DNA polymerase. Three such enzymes exist. DNA polymerases I and II (products of the genes *polA* and *polB* respectively) are not needed for

DNA replication, since mutants lacking both these enzymic activities are viable. Polymerase I is probably involved in the repair of damaged pre-existing DNA, and also in the replication of certain plasmids, such as ColE1. The function of DNA polymerase II is unknown.

DNA polymerase III, the enzyme that is used for replication, is extremely complex. At least ten different polypeptides are involved – several of them as dimers – so the holoenzyme may consist of no fewer than 22 individual polypeptide chains. In addition, the DNA duplex ahead of the replication fork is opened up by a helicase and SSB protein. A topoisomerase is also required to act ahead of the polymerase to relieve the torsional strain arising from the unwinding of the duplex. The enzyme can incorporate about 800 nucleotides per second, so that the whole *E. coli* chromosome is replicated in about 40 minutes. Since replication is bi-directional $(3{\cdot}8 \times 10^6)/2$ bp are copied as each replication fork moves round half the chromosome. RNA primer is removed from the Okazaki fragments by the action of ribonuclease H that specifically hydrolyses RNA in covalently joined RNA–DNA hybrid molecules. The missing nucleotides are filled in by DNA polymerase I and the new DNA fragments are joined by DNA ligase (Fig. 1.8).

Several forms of DNA polymerase III have been isolated and studied. Some of these have only limited function. The core enzyme, designated Pol III, contains the largest subunit α and two smaller subunits ε and θ. ε is required for proof-reading (see below), but the function of θ is not known. This form of the enzyme only incorporates about ten deoxy-nucleotides into a new DNA chain before dissociating from the template. For proper functioning in replication DNA polymerase needs a much higher degree of *processivity* – that is to say it must catalyse the incorporation of a very large number of deoxynucleotides into the new chain before dissociating from the template. Pol III′ contains an additional subunit, τ, that contributes to the processivity of the enzyme and is also

Fig. 5.1. The replication fork. It is travelling to the right. The newly synthesised DNA is shown with heavy lines. The ellipse is the core protein plus associated proteins, G is the primase and H is the helicase.

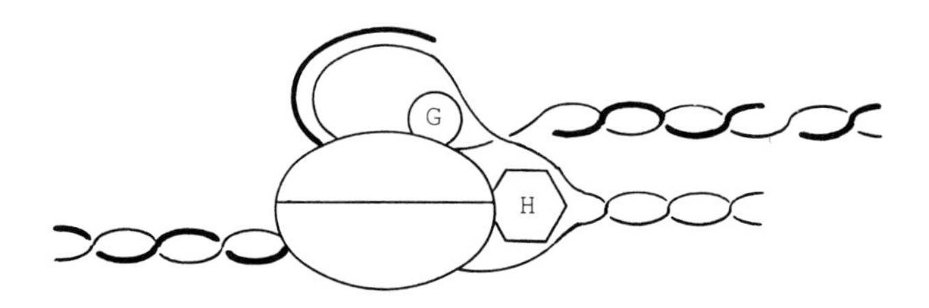

required for its dimerisation. Pol III* contains, in addition, γ and δ and is probably an artefact of preparation. Finally the holoenzyme contains the remaining subunits β, χ, and ψ. This is now highly processive.

A number of other proteins are also involved, but only the function of n′ is known. This protein is a helicase translocating processively in the $3' \rightarrow 5'$ direction, unlike the dnaB protein that translocates in the opposite direction. It is likely that n′ works on the leading strand, while dnaB protein works on the lagging strand.

It is believed that the holoenzyme has an asymmetric structure (Fig. 5.1). The smaller part works on the leading strand and the larger one on the lagging strand. If the two unequal parts of the holoenzyme do work in this way, the lagging strand must be looped round so that the polymerase can catalyse the synthesis of both strands simultaneously.

The two subunits τ and γ have an interesting relationship, since they are both products of the same gene *dnaZX*. The γ subunit comprises only the N-terminal two-thirds of the τ protein. It is an open question whether it is produced by strictly controlled proteolysis of τ or whether it arises by an unusual transcriptional mechanism involving stalling of the RNA polymerase at a rare codon on the mRNA followed by a shift in the reading frame that enables a nearby termination codon to be read and used.

The synthesis of a new DNA chain is not completely error free – about 1 nt in every 10000 is misincorporated, probably due to the occurrence of rare tautomeric forms of the bases that pair in an unusual fashion with bases on the template. This is an unacceptably high rate of error since, if uncorrected, it could give rise to the production of proteins which are either non-functional or have altered properties. In any case, estimates of the natural mutation rate are much lower than this. It is believed that newly synthesised DNA is subject to 'proof reading' – a process in which mis-paired nucleotides are excised by the $3' \rightarrow 5'$ exonuclease action of the ε subunit of DNA polymerase III, giving the polymerase the chance to incorporate the correct nucleotide before proceeding with the synthesis of the DNA. This reduces the error rate to something like 1 in 10^8. Still further reduction is achieved by a set of mis-match correction enzymes and proteins, that can detect mis-pairing of bases on the newly synthesised DNA strand and cut out a stretch of bases around the mis-match by nucleolytic action. It is believed that DNA polymerase I fills in the resulting gap reducing the overall error rate to about 1 to 10^{10}.

When growth occurs at very high rates, and cells are dividing more rapidly than once every 40 minutes, multiple replications can take place from the oriC on the same chromosome (Fig. 5.2).

5.2 Transcription

Transcription requires RNA polymerase – a large oligomeric enzyme that catalyses the formation of phosphodiester bonds between ribonucleoside triphosphates on a DNA template with the release of inorganic pyrophosphate (Fig. 5.3). The core enzyme consists of three subunits, α, β and β′ in the ratio of 2:1:1. Other polypeptides (sigma (σ)

Fig. 5.2. Multiple replication on one prokaryotic chromosome. RF are the replication forks where new DNA synthesis is occurring. RF′ are the replication forks where the second round of DNA synthesis occurs before the first round is completed. ——original strands of DNA, ---DNA strands synthesised on first round of replication, ... DNA strands synthesised on second round of replication.

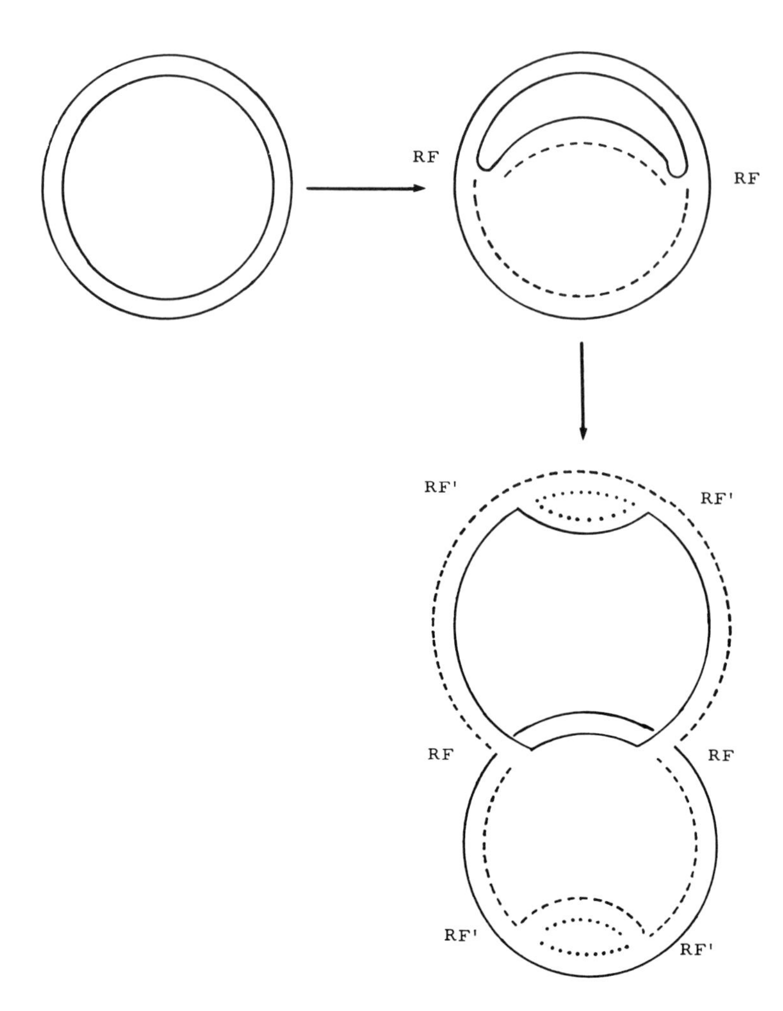

and rho (ρ)) may associate with the core enzyme temporarily during the initiation and termination of transcription. In purified systems the core enzyme binds indiscriminately to DNA and then proceeds to catalyse the synthesis of an RNA strand. However, *in vivo*, initiation of RNA synthesis starts at specific sites just before genes specifying functional units. σ promotes this specificity of binding that allows initiation to occur correctly.

Sites for this are signalled by two sequences in the DNA just upstream from the transcription start site, known as *promoters*. One, centred at 10 nt upstream of this site, is known as the Pribnow box, after its discoverer, and has the consensus sequence TATAAT: the second is about 25 nt further upstream with the consensus sequence TTGACA. This is referred to either as the −35 box or the recognition site, where RNA polymerase is believed to bind to the promoter. A crucial feature of this region is the spacing between the Pribnow box and the recognition site. For optimal rates of transcription, it should be exactly 17 nt long, but the actual sequence seems to be of much less importance. A promoter containing these two sequences separated by exactly 17 nt is more effective than several natural promoters, but strangely enough, this optimal sequence has not so far been shown to occur naturally. The precise sequences of the Pribnow box and recognition site are important in determining the amounts of transcripts that are normally made in the cell.

Fig. 5.3. Transcription, showing the site at which nucleotides are being added to the growing RNA chain. To the left, the newly synthesised RNA is dissociating from the sense strand of the DNA, which is about to reassociate with its complementary anti-sense strand. To the right, the two DNA strands are separating under the influence of RNA polymerase. The incoming nucleoside triphosphates lose pyrophosphate on being incorporated into the growing RNA chain. The region covered by RNA polymerase where the two DNA strands are separated is much larger than that shown here.

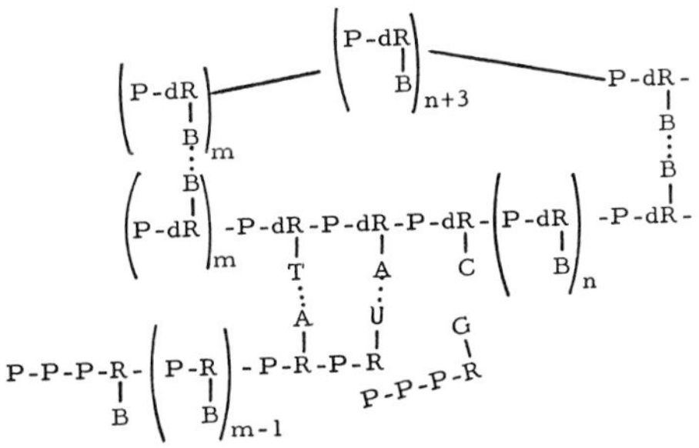

```
                      ( P-dR )
   ( P-dR )———————————(  |   )————————————————P-dR-
   (  |   )           (  B   )n+3               |
   (  B   )m                                    B
      :                                         :
   (  B   )                   (      )          B
   (  |   )                   (      )          |
   ( P-dR )m -P-dR-P-dR-P-dR- ( P-dR )  -P-dR-
                |     |     | (  |   )
                T     A     C (  B   )n
                :     :
                A     U     G
                |     |      \
P-P-P-R- ( P-R ) -P-R-P-R     P-P-P-R
         |       (  |  )
         B       (  B  )m-1
```

These sequences are often somewhat different from the consensus, with weak promoters showing greater divergence than the stronger ones. Point mutations in these sites lead to predictable increases or decreases in promoter efficiency.

E. coli has at least five different σ factors, some designated on the basis of their relative molecular masses as σ^{70}, σ^{54}, σ^{32}, σ^{E}, and σ^{F}. σ^{70} is the one that is most frequently used. σ^{32} and σ^{E} are used for transcription of heat shock protein genes (Chapter 7.10), while σ^{54} binds RNA polymerase to a special promoter site on the ALG operon that is used when *E. coli* is growing in a medium deficient in ammonium. σ^{F} is used in the transcription of genes coding for proteins involved in motility and chemotaxis. The promoters of these genes have -10 and -35 elements differing from the more usual TATAAT and TTGACA sequences associated with other genes.

Bacillus subtilis has at least six σ factors, some of which are used at different stages of sporulation.

There is some homology between all these different σ factors, and the proteins are organised into distinct domains. Two of these domains possess the helix-turn-helix motif involved in binding to DNA (Chapter 7.9), while other domains probably bind the TATA box or its equivalent, and another has homology to single strand binding proteins so that it could function to keep the DNA duplex unwound. The σ subunit dissociates from the RNA polymerase holoenzyme after the first ten nucleotides have been transcribed. It is then free to associate with another molecule of the polymerase to produce a further transcript. The maximum rate at which RNA polymerase initiates transcription is about one or two chains per second, while the rate of elongation is about 60 nt per second – an order of magnitude slower than DNA replication. Initiation generally begins with a purine nucleoside triphosphate, using ATP more commonly than GTP.

When RNA polymerase containing the σ subunit (referred to as the holoenzyme) binds to the promoter it is said to form a closed complex with the DNA base-paired in a double-stranded configuration. Its first action is to separate the two DNA strands, forming an open promoter complex. It now starts transcribing the 'sense' strand of the DNA in the $3' \rightarrow 5'$ direction. Note that the RNA is formed in the opposite $(5' \rightarrow 3')$ direction.

Termination of transcription also occurs at specific sites containing a run of dA residues in the sense strand of the DNA, preceded by a pair of inverted repeats that can form a stem and loop structure (Fig. 2.2). These stems are generally stabilised by a high proportion of G and C residues and may cause the polymerase to pause and thus aid its dissociation from the template. Dissociation of the polymerase from the DNA may also be facilitated since the rU–dA pairs that are formed by transcription from the dA sequences are exceptionally unstable.

Some transcripts require the termination factor ρ, but this is not a universal requirement. Where it is used, the actual base sequence at the termination site does not seem to be well conserved. ρ binds to sites of about 80 bp some way upstream from the actual termination site. The protein is a hexamer that possesses a distinct RNA binding site and also has ATPase activity. However, the molecular details of its involvement in termination of transcription are not yet known.

During trancription the newly formed RNA dissociates fairly rapidly from the complementary DNA strand as the polymerase passes along, and the double-stranded DNA re-forms behind the enzyme.

The majority of bacterial genes are transcribed into mRNA. mRNAs contain a ribosome-binding site a few bases upstream from the initiation codon which is nearly always AUG. This is known as the Shine–Delgarno sequence, and has the consensus sequence AGGAGGU. It is complementary to seven bases at the 3′-end of the 16S rRNA (Fig. 5.4), and plays an important role in positioning the mRNA correctly on the ribosome.

Translation of the mRNA starts while it still growing by transcription from the DNA. In prokaryotes mRNA is generally very unstable, being rapidly hydrolysed by abundant ribonucleases that are present in the cell, so that its average half-life is only 2–4 minutes.

5.3 Some RNAs are processed after transcription

Some of the RNA that is synthesised by transcription is used directly, either as tRNA or as rRNA. In *E. coli* there are seven genes coding for long RNA molecules which contain sequences of all three of the different rRNA species (Table 2.1), as well as some kinds of tRNA (Table 5.1) The sequences of the rRNAs encoded by these genes are all

Fig. 5.4. Interaction of Shine–Delgarno sequence on prokaryotic mRNA with the 3′-end of 16S rRNA. A similar interaction is believed to occur when eukaryotic mRNA is positioned on the ribosome.

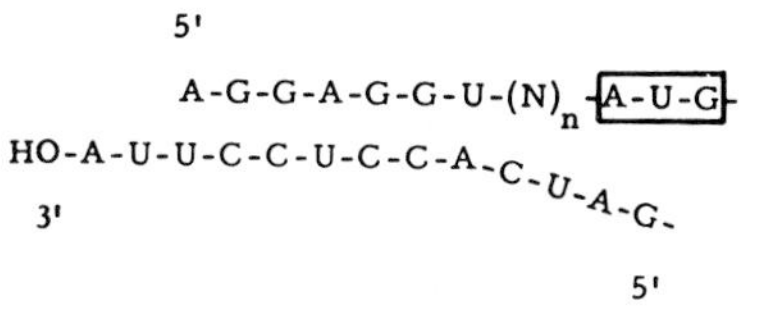

```
        5'
        A-G-G-A-G-G-U-(N)n-[A-U-G]-
HO-A-U-U-C-C-U-C-C-A-C-U-A-G-
3'                              5'
```

Table 5.1. *Ribosomal RNA operons in* E. coli

Symbol	Map position (min)	Order of genes present					
rrnA	86	rrsA	alaT	ileT	rrlA	rrfA	
rrnB	89	rrsB	gltT	rrlB	rrfB		
rrnC	84	rrsC	gltU	rrlC	rrfC	aspT	trpT
rrnD	72	rrsD	alaU	ileU	rrlD	rrfD	thrV
rrnE	90	rrsE	gltV	rrlE	rrfE		
rrnG	56	rrsG	gltW	rrlG	rrfG		
rrnH	5	rrsH	alaV	ileV	rrlH	rrfH	aspU

rrs, rrl, and rrf stand for 16S, 23S and 5S rRNA respectively.
The other genes code for tRNAs for the amino acids indicated by their standard abbreviations.

identical. The transcripts of three of them contain the tRNAs for isoleucine and alanine in part of the spacer region between the 16S and 23S RNAs, while the other four transcripts contain a tRNA for glutamate in that position. In addition, one gene cluster of each type contains a gene for a tRNA for aspartate downstream from sequences encoding the rRNAs. This suggests that multiple rRNA gene clusters have arisen by duplications from a common ancestor which has had different tRNA genes inserted at various times during the course of evolution.

The transcripts of the rRNA genes are processed by hydrolysis of phosphodiester bonds at specific sites by RNase III in *E. coli*. The rRNA precursor exhibits interesting secondary structure in which the sequences that form the mature 16S and 23S molecules are looped out at the end of fairly long base-paired stems. RNase III acts on these stems making slightly staggered cuts. There is no obvious sequence homology immediately surrounding these sites of cleavage. Further trimming of a comparatively small number of nucleotides at both ends of the immediate precursors of these rRNAs follows (Fig. 5.5).

The formation of the rRNA leaves some tRNAs embedded in sequences of nucleotides that must be removed to generate the mature molecules. Other tRNAs are also transcribed as larger precursors – some individually; others in small clusters of up to four. These are all processed by RNase P, an endonuclease that cleaves the precursors at the junction with the 5′-end of the tRNA, and RNase D, an exonuclease that hydrolyses phosphodiester bonds from the 3′-end of the precursors until it reaches the 3′-terminal trinucleotide CCA where its action stops. This CCA terminus, found on all tRNAs, is essential for their function since it is the site of attachment of the amino acyl residue.

Further changes are required before the tRNAs are completely mature,

Table 5.2. *Antibiotic resistance genes carried by some transposons in* E.coli

Transposon	Size (bp)	Resistant to
Tn 3	4957	Ampicillin
Tn 5	5400	Kanamycin
Tn 9	2638	Chloramphenicol
Tn 10	9300	Tetracyclin

Fig. 5.5. (a) RNA transcript of rRNA gene cluster with two tRNAs in the spacer between the 16S and 23S rRNAs. (b) details of the secondary structure of the tRNAs and surrounding nucleotides before processing. Arrows show some of the places where processing is believed to occur. III is ribonuclease III; P is ribonuclease P.

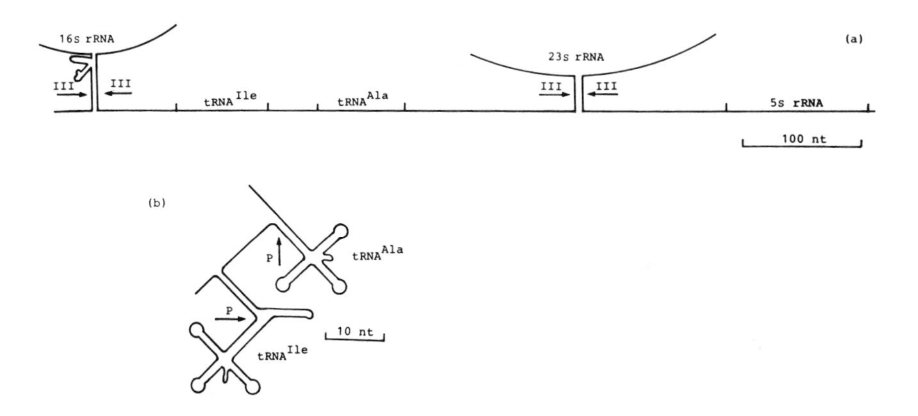

Fig. 5.6. Integration of IS1 into the *E. coli* chromosome generates a direct repeat of 9 bases at each end of the inserted element as a result of a staggered cut in the genomic DNA. This example shows insertion into the *lac*I gene.

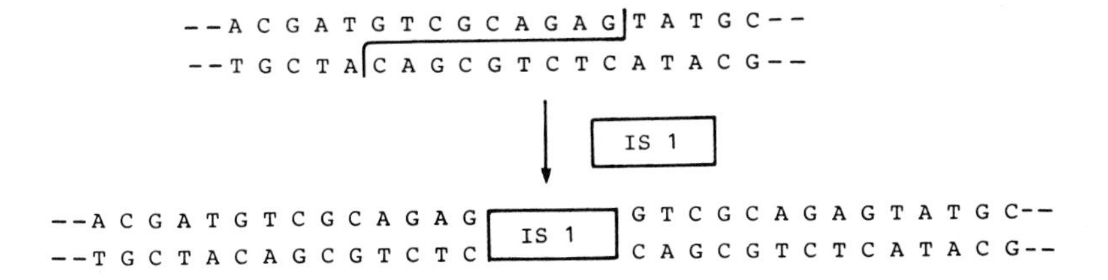

since they all contain bases that are modified from the A, G, C and U in the primary transcript (Fig. 2.4). These changes involve a number of different enzymes, catalysing a wide range of individual reactions.

5.4 Transposable genetic elements

Bacterial genomes carry multiple copies of insertion sequences (IS) and transposons (Tns) at various sites. They can move from one site to another, though often only at low frequency. There is no absolute specificity of DNA sequence at the sites at which they are inserted though certain ones are favoured over others. IS are shorter than Tns and probably carry a sequence coding for the enzyme transposase that is required for insertion to take place. IS contain regions of inverted repeats at each end. Tns contain similar regions, but these tend to be much longer, and are often IS themselves. In addition, Tns contain genes encoding enzymes conferring resistance to certain antibiotics (Table 5.2). Transcription and/or translation of genes in the region where IS or Tns are inserted is often interrupted. Less frequently, transcription may be activated since they contain promoter sequences required for transcription of their own DNA.

On insertion of IS and Tns there is a duplication of target DNA for a short length (commonly 4–9 bp) on each side of the insertion, suggesting that this event involves a staggered break in the host DNA (Fig. 5.6).

Since these are movable elements they can also be excised from the genome, thus restoring any damaged functions. However, excision is frequently inexact, so reversion to the previous state occurs rather rarely.

6
The operon concept

6.1 Genes for sets of metabolically related enzymes are transcribed as one long message

In bacteria, genes specifying enzymes which are all part of a single metabolic pathway are commonly transcribed from adjacent lengths of DNA with only short non-coding stretches between them. Such 'super-genes' are known as *operons*, and are generally transcribed as single units giving rise to *polycistronic mRNAs*. Since translation immediately follows transcription this results in the rapid production of several functionally related enzymes in equivalent amounts – a process known as *co-ordinate control*. Some of these operons and the enzymes transcribed from them are synthesised more or less constantly and are known as *constitutive*; others are subject to precise control, signalled by the presence or absence of actual or potential metabolites in the bacterial cell and these are known as *inducible* when they are switched on by a particular metabolite, or *repressible* when they are switched off by a particular metabolite.

Much of our knowledge about operons has come from experiments on mutants in which they are not functioning normally. These have allowed the identification of two classes of genes. *Structural genes* encode information for either stable RNAs or mRNAs (see Chapter 2.3), while *regulatory genes* do not themselves give rise to any products, but are immediately involved in the regulation of transcription of structural genes.

These regulatory genes are situated immediately upstream from the operon whose activity they control. In addition to the promoter (Chapter 5.2), there is another region known as the *operator* which is either directly adjacent to the promoter, or even overlaps it, and there may also be other regions involved in regulatory control.

Inducible operons are said to be negatively controlled when they are normally not transcribed due to the binding of a specific *repressor* (always, as far as is known, a protein) to the operator. Induction occurs when an *inducer* (a small molecule) binds to the repressor, altering its conformation so that it now dissociates from the operator and allows transcription to proceed.

Positively controlled inducible operons are not normally transcribed. They become active when a *co-activator* (small molecule) binds to an *apo-activator* (protein) altering its conformation in such a way as to permit it to bind to a site near the operon so that transcription is initiated.

Negatively controlled repressible operons are normally transcribed, but when a co-repressor (small molecule) binds to the apo-repressor (protein) this complex then attaches to the operator resulting in inhibition of transcription.

It is theoretically possible to envisage positively controlled repressible operons in which an activator protein is normally bound to the operator, allowing transcription to occur. Interaction with an inhibitor (small molecule) would cause the activator to dissociate from the operator so that transcription would be inhibited. However, no example of this last situation is known. The various possibilities are shown in Table 6.1. The first two modes of regulation are not mutually exclusive, and several examples are known in which control is exerted in both these ways.

The operator is generally situated between the promoter and the site at which transcription starts. It is believed that binding of a repressor causes a physical block to the passage of RNA polymerase from the promoter to the site at which it would otherwise start transcription. Repressors are encoded by structural genes which may or may not be adjacent to the operon they control, and they are generally present in very few copies per cell.

A further method of regulation is seen in a number of operons coding for enzymes required for the synthesis of amino acids and other essential metabolites. This is called *attenuation*, and is discussed later (Chapter 6.7).

As a general rule, repressible operons code for enzymes required for the synthesis of various small molecules, such as amino acids, purines and pyrimidines which are needed for other synthetic processes in the cell. These may be present in widely varying concentrations in the milieu in which the bacteria are living. When they are present in adequate amounts, considerable economy results from switching off the synthesis of the enzymes that would be required to make them. Conversely, inducible operons code for catabolic enzymes for metabolising certain sugars and amino acids which may sometimes be present in excess and can be used as energy sources. There is obviously no point in making these enzymes when

Table 6.1. *Type of control of operons*

Type of operon	State of regulatory molecules	Binding to operator	Transcription	Examples
Negative inducible	Repressor free	+	−	*lac*
	Repressor binds to inducer	−	+	*gal*
Positive inducible	Apo-activator free	−	−	*ara*
	Apo-acivator binds to co-activator	+	+	*mal*
Negative repressible	Apo-repressor free	−	+	*trp*
	Apo-repressor binds to co-repressor	+	−	*arg*

Table 6.2 Trans-*acting mutations of the genes encoding regulatory proteins*

Regulatory protein	Function inhibited by mutation	Result
Repressor	DNA binding	Constitutive expression of operon
Repressor	Inducer binding	Operon becomes non-inducible
Apo-activator	DNA binding	Operon becomes non-inducible
Apo-activator	Co-activator binding	Operon becomes non-inducible
Apo-repressor	DNA binding	Operon becomes non-repressible
Apo-repressor	Co-repressor binding	Operon becomes non-repressible

such substrates are not available and the bacterium is utilising more generally available energy sources such as glucose.

Much of the information about the regulation of operons has been gained from studies of mutations mapping in the various regulatory regions. Mutations in the gene encoding repressor, apo-repressor or apo-activator can have various effects, depending on the site of the mutation. They can affect the interaction of the protein product with the operator. Thus, an altered repressor may no longer bind to the operator, in which case a negatively controlled inducible operon can become constitutively expressed. Alternatively, a mutation which causes defective binding of an inducer by the repressor will lead to an operon never being expressed. These and other possibilities are presented in Table 6.2. Mutations in the operator can obviously affect the binding of repressors and other regulatory proteins, as shown in Table 6.3.

Mutations in genes coding for regulatory proteins are said to act in *trans* because the products that they supply can act on chromosomes other

Table 6.3. cis-*acting mutations of the operator*

Negative inducible	Inhibits binding of repressor	Constitutive expression of the operon
Positive inducible	Inhibits binding of apo-activator + co-activator	Operon is not expressed
Negative repressible	Inhibits binding of apo-repressor + co-repressor	Loss of ability to repress operon

than the one in which they are found. Operator and promoter mutations are said to act in *cis* because they will only affect the functions of the chromosomes on which they are sited. *trans*-acting mutations can be corrected (complemented) by the introduction of a plasmid bearing the un-mutated gene, while *cis*-acting mutations cannot be corrected in this way.

A common feature of many inducible operons which are positively controlled is that the induction is caused by the binding of a protein known alternatively as Catabolite Activator Protein (CAP) or Cyclic-AMP Receptor Protein (CRP). This protein, as its alternative name implies, also has a binding site for cAMP, and it will interact only with regulatory sites on DNA after it has bound this nucleotide. These regulatory sites are generally situated just upstream from the RNA polymerase recognition site (the −35 box), and it is suggested that the binding of CAP makes it easier for the polymerase to bind here. A feature of some of these sites is that there is a region of symmetry in the DNA sequence extending over about 16 bp (Fig. 6.1). This may be important in the recognition of appropriate DNA sequences by CAP, since this is a dimer of two identical subunits.

Mutants which are defective in the gene coding for CAP (*crp*) or in that encoding adenylate cyclase (*cyc*) fail to show induction of operons that are normally induced by CAP.

Fig. 6.1. Region of inverted symmetry at sites where CAP is believed to bind in the *lac* (top) and *gal* (bottom) operons. Lines are drawn between the base sequences which exhibit the symmetry, and large dots mark the centres of symmetry. Note that stems giving rise to cruciform structures can be formed between the over- or under-lined bases in one strand.

G T G A G T T A G C T C A C
C A C T C A A•T C G A G T G

G T G T A A A C G A T T C C A C
C A C A T T T G•C T A A G G T G

Table 6.4. *Some examples of the control factors for operons*

Operon	Inducer	Co-activator	Co-repressor	Type of operon
lac	isopropyl-thio-galactoside allo-lactose	cAMP	—	Negative inducible / Positive inducible
gal	galactose	cAMP	—	Positive inducible / Negative inducible
mal	maltose	cAMP	—	Positive inducible / Negative inducible
hut	urocanate	cAMP	—	Positive inducible
trp	—	—	tryptophan	Negative repressible
arg	—	—	arginine	Negative repressible

Adenylate cyclase is stimulated when the concentration of glucose falls to a low level. This leads to the synthesis of cAMP which is then bound to CAP making it possible for this protein to bind to sites on inducible operons which provide enzymes for the catabolism of alternative energy sources (e.g. lactose, galactose, arabinose, histidine).

The foregoing description gives a general picture of some of the features that are frequently found in operons, but there are many individual variations. Several operons are discussed in more detail in the following sections so as to exemplify some of these differences (Table 6.4).

6.2 The *lac* operon

This was the first operon to be discovered, and it has been intensively studied over the past 20 years. It consists of three structural genes coding for proteins involved in the uptake and catabolism of lactose (Fig. 6.2). Upstream of the structural genes are an operator, a promoter and a CAP-binding site, and also the structural gene for the *lac* repressor protein. The operon is normally induced when lactose appears in the medium and when the concentration of glucose (a preferred energy source) drops to a low level. It is controlled in two ways – negatively by the *lac* repressor protein which dissociates from the operator when lactose becomes available, and also positively by CAP.

The operator site where the *lac* repressor binds is mostly in the region 5′ to the initiation codon and extends just 5′ to the site where transcription starts. It contains a 35 bp sequence with a high degree of inverted symmetry (only seven bases do not match). This symmetrical arrangement may facilitate the binding of the *lac* repressor which is a tetramer of four identical subunits. In fact, a completely symmetrical operator has been synthesised and shown to bind the *lac* repressor with an affinity ten-fold greater than the natural operator.

The gene for the *lac* repressor has its termination codon at or very near the 5′-end of the CAP-binding site. The gene is preceded by very weak promoter sequences and therefore is poorly transcribed. It has been estimated that there are only about ten molecules of *lac* repressor in a cell of *E. coli.* However, because of the high affinity it displays for the *lac* operator, this is a sufficient quantity to keep the *lac* operon repressed in the absence of lactose. In order to obtain enough of this interesting protein for study, the gene has been engineered into a plasmid immediately downstream from a strong promoter. When the plasmid is grown in a suitable host large quantities of the repressor are produced.

The actual inducer of the *lac* operon is not lactose, but an isomer of this sugar called allo-lactose (Fig. 6.3). Its synthesis from lactose is catalysed by the first enzyme specified by the operon – β-galactosidase – and, like lactose, it can also be hydrolysed by this enzyme. Some other, non-metabolisable inducers have been synthesised, such as isopropyl-β-thio-galactoside (Fig. 6.3). Such inducers are called gratuitous inducers.

6.3 The *gal* operon

The *gal* operon encodes three structural genes for the catabolism of galactose (Fig. 6.4), and is derepressed by interaction of free galactose

Fig. 6.2. The *lac* operon. Top: Map of the complete operon. I, Z, Y, A are the symbols for the structural genes with the names of the proteins that they encode given below the operon. c is the control region. Bottom: Detail of the control region. The numbers refer to the conventional numbering of nucleotides, beginning at +1 at the 5′-most transcribed nucleotide. Numbering is negative from this point into the 5′-flanking region.

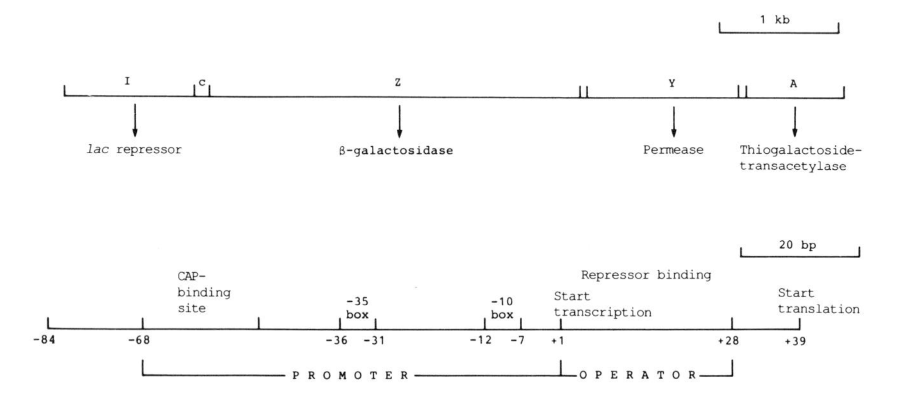

with the *gal* repressor protein. However, it has some features which distinguish its mode of control from that of the *lac* operon. An operator site was originally identified 5′ to the RNA polymerase-binding site, and therefore it did not seem possible that when the repressor was bound it could block transcription. However, a second operator has recently been discovered which is actually situated within the first structural gene. It is believed that the dimeric *gal* repressor protein binds to each operator through one of its monomers, thus effectively preventing transcription of the operon (Fig. 6.5).

Fig. 6.3. Structure of lactose, the physiological inducer of the *lac* operon; allo-lactose, the actual inducer, produced by enzyme action on lactose; β-isopropyl-thiogalactoside, a synthetic gratuitous inducer.

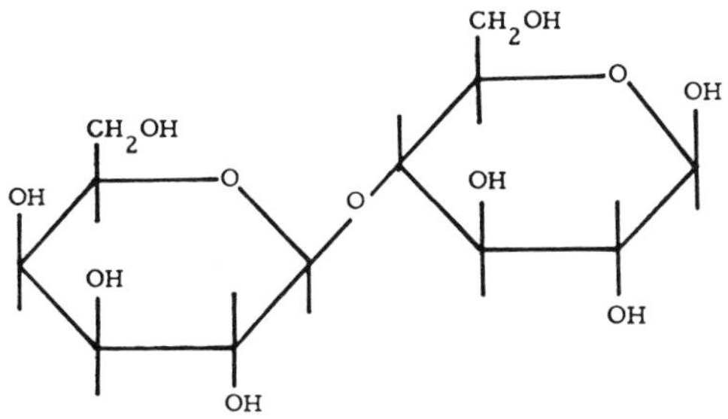

Lactose

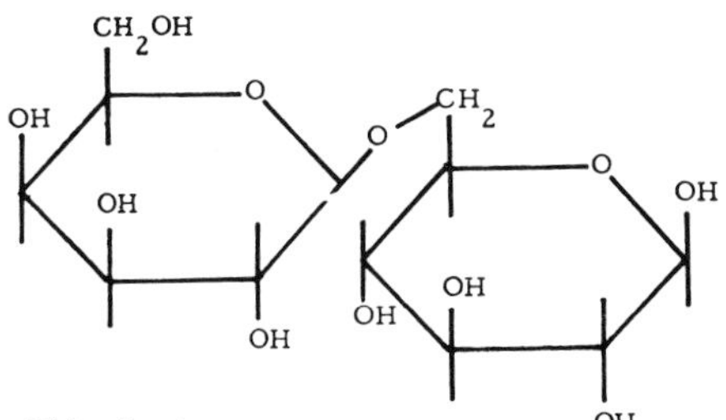

Allo-lactose

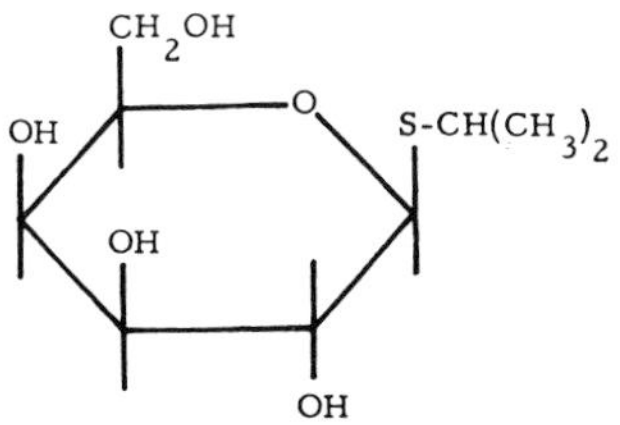

β-Isopropyl-thiogalactoside

Another interesting feature of this operon is that there are two promoter sites, separated by 5 bp (Fig. 6.6). The Pribnow boxes overlap

Fig. 6.4. The *gal* operon. Top: Map of complete operon. Bottom: Details of the control region. For conventions, symbols and numbering see Fig. 6.2.

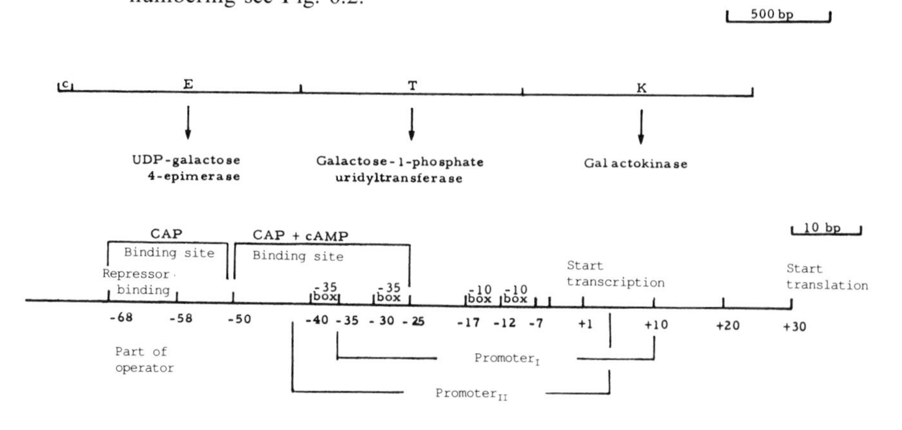

Fig. 6.5. The operator of the *gal* operon. The *gal* repressor binds at both the sequences shown, which are situated at approximately equal distances upstream and downstream from the initiation codon, which is shown on the right.

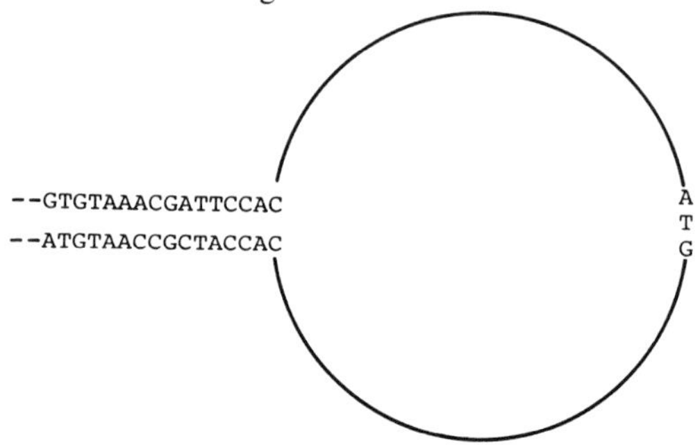

Fig. 6.6. The two promoters of the *gal* operon. P_1 is activated in the presence of CAP-cAMP which leads to transcription starting at the A residue numbered +1. Its −10 and −35 boxes are outlined in full lines. P_2 directs transcription at the A residue numbered −5, and is used in the absence of CAP-cAMP. Its −10 and −35 boxes are shown outlined in dashes.

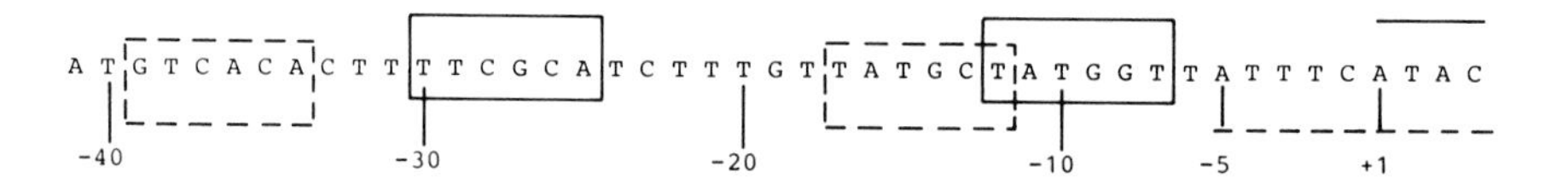

slightly, and are used to direct the RNA polymerase to start transcription at one or other of the alternative sites. There are also adjacent −35 boxes. The 3′-most promoter site is known as P_1 and transcription using this promoter is stimulated by CAP binding at a site overlapping the −35 box. There is a second binding site for CAP to the 5′-side of the first. Binding of the second molecule is facilitated by the prior binding of RNA polymerase. CAP binding inhibits transcription from the other promoter site (P_2), which is used only in the absence of CAP. CAP-independent transcription of the operon occurs at a low rate normally because the first two enzymes encoded by the operon are required for the production of galactose from glucose, as well as for the catabolism of galactose. The former process is essential, since small quantities of galactose are needed for the synthesis of some cell wall constituents of *E. coli*. The expression of the operon shows a marked degree of *polarity* – that is to say the distal enzyme (galactokinase, which is needed only for the catabolism of galactose – not for its production from glucose) is produced in considerably smaller amounts than the two proximal enzymes (the epimerase and transferase). On full induction of the operon, in the presence of CAP-cAMP, production of the kinase is stimulated to a greater extent than that of the other two enzymes.

A further difference between the *lac* and *gal* operons is the site of the genes coding for their repressors. That for the *gal* operon is situated on the opposite side of the *E. coli* chromosome.

6.4 The *ara* operon

Once again, an operon controlling the synthesis of catabolic enzymes (this time for the sugar arabinose) contains three structural genes (*ara*B, A and D) (Fig. 6.7). In this case the gene for the repressor (*ara*C) is very close to the structural genes. It is separated from them by only 147 bp, and it is transcribed from the opposite strand of the DNA, and therefore in the opposite direction. The 147 bp stretch contains all the regulatory sites. The *ara* repressor binds to the operator just 5′ to its own gene and represses its own transcription, so this is an example of autoregulation. However, in the presence of arabinose, the sugar combines with the repressor and the complex binds to a second operator (the BAD operator) 5′ to the structural genes, and stimulates their transcription. Transcription of both sets of genes is also stimulated by the binding of CAP–cAMP at a single site between the two operators. The BAD genes are transcribed when RNA polymerase binds between their promoter (P_{BAD}) and the CAP-binding site. The enzyme also binds in a second place between the other end of this site and the promoter site for the repressor gene. This second site overlaps the operator for the C gene. Each promoter contains a Pribnow box and a −35 box. From their sequences it is predicted that P_C is a fairly strong (though not ideal) promoter, while P_{BAD} is a poor one.

In the presence of arabinose there is probably a small amount of transcription of *ara*C, especially as the *ara* repressor binds to the O_c site about ten times less strongly than the complex of repressor + arabinose.

In summary, there is positive control of P_{BAD} by CAP and by the complex between arabinose and its repressor. P_C is also positively controlled by CAP, but negatively controlled by the repressor especially when it is complexed with arabinose.

6.5 The *hut* operons

The catabolism of some amino acids is also inducible with negative control by a repressor protein. An example of this is found in the *hut* operons (histidine utilisation) which are switched on when there is a plentiful supply of histidine in the medium. Excess histidine can be degraded by bacteria to provide energy. This system has been most studied in *Salmonella typhii*.

The four genes required for this catabolic pathway are found in two

Fig. 6.7. The *ara* operon. Top: Map of the complete operon. For conventions about symbols and numbering, see Fig. 6.2. The horizontal arrows show the directions of transcription. Bottom: Detail of the control region. Numbers above the line are with reference to the transcription start site of the *ara* BAD gene cluster: numbers below the line are with reference to the transcription start site of the C gene.

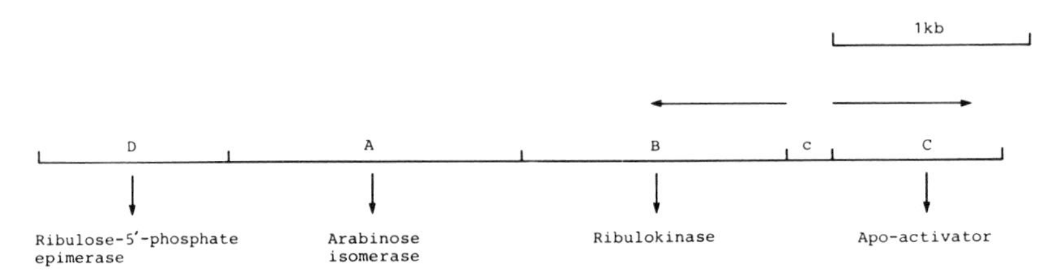

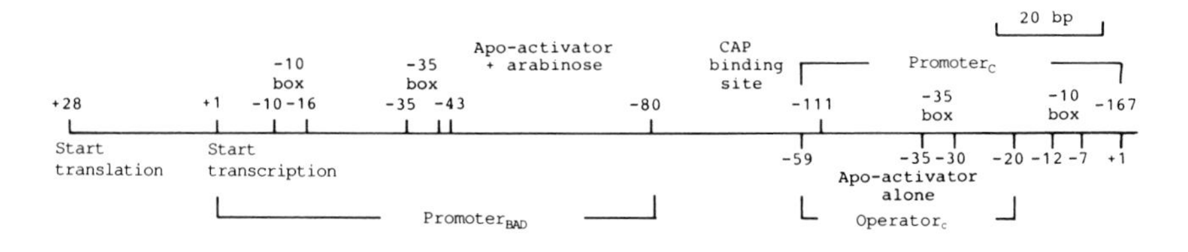

operons which are adjacent; the left-hand one also contains the gene for the repressor (Fig. 6.8). This binds to operators designated m and p which are situated immediately 5′ to the left-hand and right-hand operons respectively. It binds less strongly to the m operator so that there is presumably some read-through of the left-hand operon in the presence of low concentrations of repressor, until a certain level of this protein is reached and further synthesis is inhibited.

The inducer is not histidine itself, but the first metabolite to be produced during its catabolism – urocanate – so there must be small amounts of histidine and the enzyme histidase available for the operon to be induced.

Catabolite repression is also involved in the control of these operons since CAP-cAMP stimulates transcription of the right-hand operon.

6.6 The *mal* regulon

The *mal* operons, coding for proteins involved in the uptake of maltose into the cell and its catabolism, exhibit a further degree of complexity. There are two separate operons, located at different sites on the *E. coli* chromosome, which are under the common control of a single apo-activator (Fig. 6.9). Such a functionally linked series of operons is known as a *regulon*.

The *mal*A region contains the genes *mal*P and *mal*Q coding for enzymes involved in the catabolism of maltose, and also the *mal*T gene which codes for the apo-activator of this system. *mal*T is transcribed in the opposite direction to *mal*P and *mal*Q, in a manner reminiscent of the repressor and

Fig. 6.8. Schematic diagram of the *hut* operon. The precise details are not yet known.

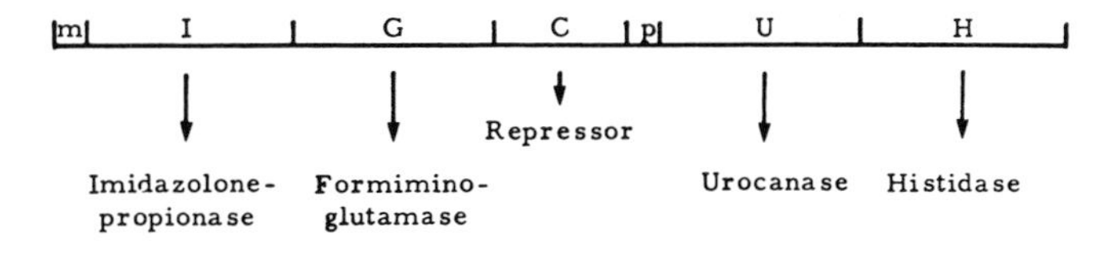

Fig. 6.9. Schematic diagram of the *mal* regulon. The A and B regions map to 74 min and 90 min respectively on the *E. coli* chromosome.

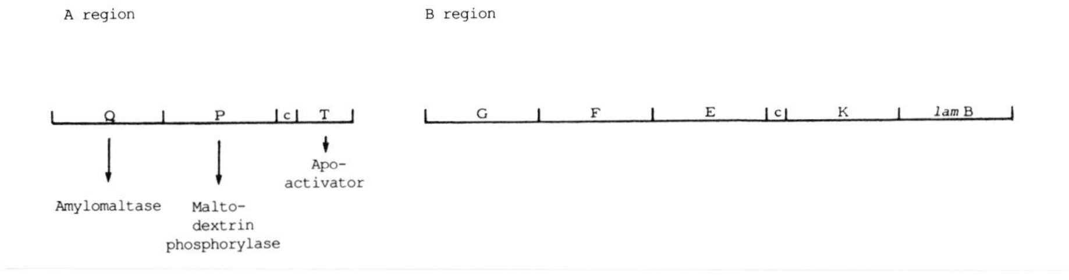

structural genes of the *ara* operon. The *mal*B region encodes five structural genes. These are arranged in two transcription units which are transcribed in opposite directions with the interaction site(s) for the *mal*T gene product situated between them. The products of all genes are involved in the uptake of maltose into the cell. The *lam*B gene product was originally shown to be required as a receptor for the uptake of the bacteriophage λ into the cell and only later found to stimulate the uptake of low concentrations of maltose.

The transcription of *mal*T is strongly stimulated by CAP-cAMP. After binding maltose, the *mal*T gene product activates transcription of *mal*P and *mal*Q, probably independently of CAP-cAMP. However, the transcription of the genes in the *mal*B region requires the binding of both CAP-cAMP and the apo-activator plus maltose. Thus a major physiological effect of the production of cAMP, which will occur when glucose concentration falls, is to allow entry of maltose into the cell in large amounts and this enhances the transcription of the genes encoding the enzymes required for its catabolism.

6.7 The *trp* operon; control by attenuation

There are a number of operons containing the structural genes coding for the enzymes required for the synthesis of amino acids. These are repressed, either by the amino acids themselves or by easily formed derivatives, so that unnecessary, energetically wasteful, synthesis of these compounds does not occur when there are adequate amounts available to the cell.

The *trp* operon encodes five structural genes required for the synthesis of tryptophan from chorismate (Fig. 6.10). An unlinked apo-repressor is encoded by the *trp*R gene, and when this combines with tryptophan the complex binds to an operator site just 5′ to the site at which transcription starts. This inhibits the passage of RNA polymerase from the promoter and so turns off transcription.

There is a sequence of 162 bp between the start of transcription and the

Fig. 6.10. The *trp* operon. In addition to the structural genes, the position of two terminators (t_1 and t_2) are shown. t_1 is a rho-independent terminator, while t_2 functions only in the presence of rho.

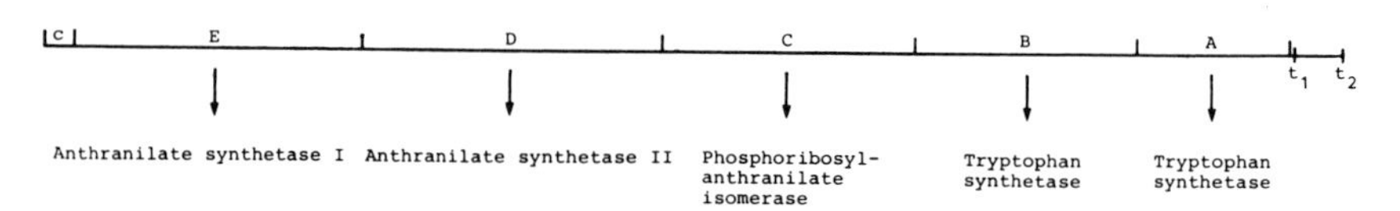

codon at which translation of the structural genes begins, referred to as the leader sequence. It is involved in a second method of control of this operon. The RNA transcribed from this leader sequence contains a short open reading frame of 42 bp, in which there are two consecutive codons for tryptophan. It ends with a typical termination signal – a region rich in G–C base pairs that can form a stem–loop structure, followed by a polyU tract (Chapter 5.3). This RNA transcript can form two alternative stable secondary structures by base pairing that are mutually exclusive, because part of one stem is common to both (Fig. 6.11). In the absence of other factors, structure I forms and permits the formation of the terminator stem and loop. *In vivo*, a ribosome associated with the newly transcribed RNA catalyses the synthesis of the short polypeptide encoded by the leader sequence. When it reaches the termination codon UGA at position 68, the ribosome overlaps it enough so that the structure III is not free to form. The RNA polymerase is ahead of the ribosome and transcribes through so that the structure II can form, causing it to stop transcription before reaching the AUG codon at position 163. When there is a deficiency of tryptophan and hence of charged $tRNA^{Trp}$, the ribosome stalls at one of the tryptophan codons. This allows structure III to arise so that formation of the terminator stem and loop is pre-empted. In this situation the RNA polymerase is about 150 nucleotides beyond the point where the RNA transcript emerges from the ribosome, and it reads through the potential terminator site and continues transcribing the operon. The various sequences on the RNA that form the stem structures are known as the terminator, the pre-emptor (which includes part of one limb of the terminator), and a third one, nearer the 5′-end of the RNA, called the protector that can protect the terminator by forming a stem with one limb of the pre-emptor.

This form of control, dependent on a mechanism which is responsive to the concentration of a particular charged amino acyl-tRNA in the cell is known as attenuation, and the region in which the RNA is terminated when the operon is repressed is called the *attenuator*. Both this system and the binding of the repressor to the operator seem to be operative in vivo for control of the tryptophan operon. Binding of the repressor causes a 70-fold reduction in transcription of the operon, while attenuation leads to an extra 10-fold reduction.

The biosynthesis of several other amino acids is under very similar control with leader DNA sequences that encode short polypeptides containing multiple codons (usually adjacent) for the amino acid whose synthesis the operon controls. The RNAs transcribed from these sequences are all capable of base pairing to form protector, pre-emptor and

Fig. 6.11. The control region of the *trp* operon, showing the formation of alternative stem and loop structures on the mRNA transcribed from it. Structure I can form *in vitro* in the absence of other factors. Structure II forms in the presence of a ribosome which is actively translating the mRNA for the leader peptide. The formation of a terminator stem and loop proceeds and blocks further transcription of the operon by RNA polymerase. Structure III forms in the presence of a ribosome which has paused at the *Trp* codons when there is an absence of charged $tRNA^{Trp}$. The two *Trp* codons (UGG) and the termination codon (UGA) in the leader sequence, and the initiation codon (AUG) of the *Trp* E gene are shown. Some bases are numbered from the beginning of the leader sequence.

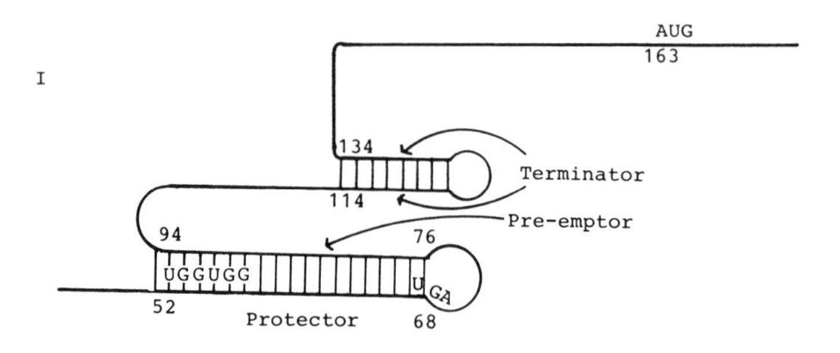

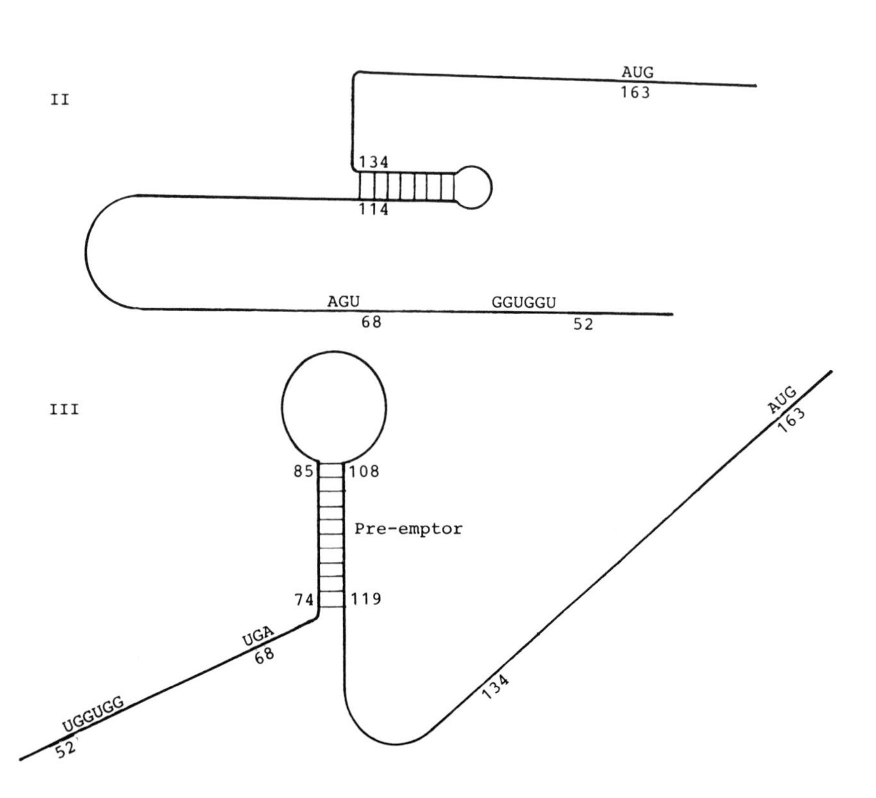

terminator elements. Amino acids whose biosynthesis is regulated in this way include phenylalanine, threonine, isoleucine, leucine and valine (in *E. coli*), and histidine and leucine in *Salmonella typhii*.

Mutation of the genes coding for the amino acid-tRNA synthetases or the tRNAs generally leads to permanent induction of these biosynthetic operons, since there is likely to be a lack of the necessary charged tRNA so that the ribosomes will be permanently stalled on the leader sequence.

6.8 Pyrimidine biosynthesis

The whole pathway for the biosynthesis of pyrimidines requires six enzymes whose genes are dispersed over the *E. coli* chromosome and constitute another example of a regulon.

Attenuation is involved in the regulation of the transcription of the enzyme aspartate transcarbamylase which is the first enzyme in this pathway. The gene encoding it has a leader sequence with an open reading frame at the end of which is a typical terminator with a potential stem and loop structure followed by a run of A residues. Earlier in the leader sequence is a second run of A residues (which will be transcribed into U residues on the RNA) and it is believed that, in the presence of low concentrations of UTP, RNA polymerase will pause at this site. The ribosome translating the mRNA catches up with the pausing RNA polymerase and when this enzyme moves forward to transcribe the region of the terminator, hairpin formation is precluded by the adjacent ribosome so that read through into the structural gene for aspartate transcarbamylase occurs. In the presence of large amounts of UTP the ribosome is further away and hairpin formation occurs at the attenuator so that transcription is terminated (Fig. 6.12).

6.9 Arginine biosynthesis

The biosynthesis of arginine requires at least nine different enzymes, two of which are oligomeric, so a total of eleven genes is needed. There is one cluster of four, though one of them is transcribed in the opposite direction to the other three, and a cluster of two, whilst the rest are scattered over a wide expanse of the *E. coli* chromosome (Fig. 6.13). There is also a separated apo-repressor gene (*arg*R). This set of genes constitutes a regulon under the control of the cellular level of arginine which acts as a co-repressor. However, the effectiveness with which the arginine–apo-repressor complex inhibits transcription of the different genes varies considerably (Table 6.5). Note that the transcription of *arg*R is also inhibited – another example of autoregulation. Carbamyl phosphate synthetase also provides carbamyl phosphate for the synthesis of pyrimidines, and expression of the two genes encoding this enzyme is additionally subject to inhibition by uracil nucleotides, presumably acting in concert with another apo-repressor.

There are reasonably well conserved sequences in the operator regions of five of these genes (including the *arg*ECBH and the *car*AB clusters)

Fig. 6.12. Regulation of the transcription of the gene encoding aspartate transcarbamylase. Top: In the presence of a low concentration of UTP the RNA polymerase (hexagon) has paused at the run of U residues in the leader sequence (vertical bars). The ribosome (circle) translating the leader has caught up with the polymerase and prevents formation of the terminator stem and loop (short vertical bars). Transcription continues into the aspartate transcarbamylase gene. (T = termination codon of leader; I = initiation codon of the structural gene.) Bottom: In the presence of a higher concentration of UTP, the ribosome does not catch up with the RNA polymerase and the terminator stem and loop can form so that transcription is halted.

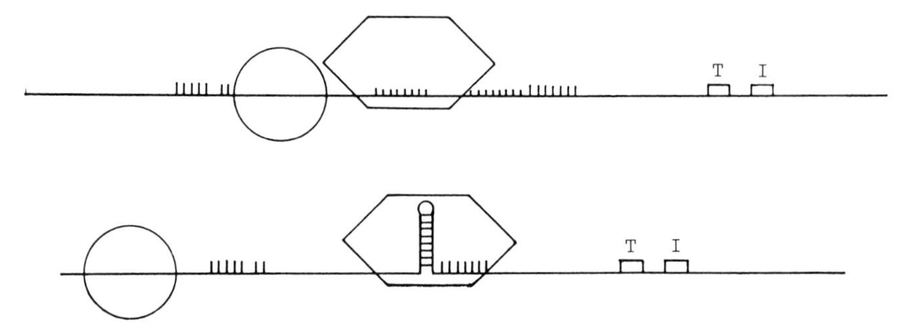

Fig. 6.13. Map of the genes of the *Arg* regulon on the *E. coli* chromosome. The chromosome is conventionally numbered in a clockwise direction from 0 to 100, starting at the top centre of this diagram. Arrows show the direction of transcription of some of the genes.

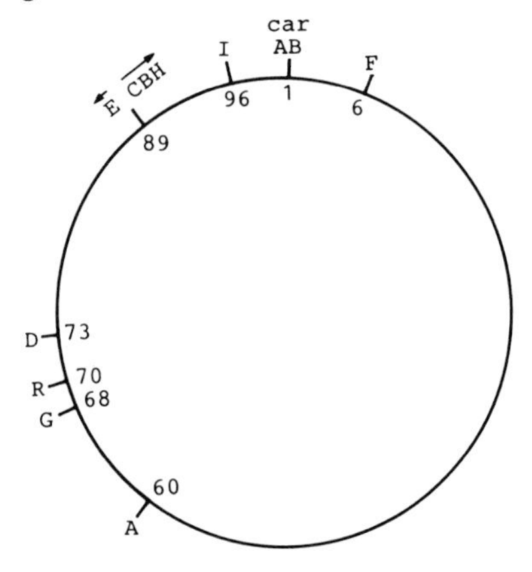

Table 6.5. *The repressibility of genes in the arginine regulon by combination of arginine with the apo-repressor*

Gene	Repressibility	
A	250	
CBH	60	
D	20	
E	17	
F	200–250	
I	400–500	
R	10	autoregulation
*car*AB	4	
	30	in presence of pyrimidines as well

Repressibility is defined as the ratio of transcription in the absence of arginine to that in the presence of 0·575 mM arginine.

which are excellent candidates for binding sites for the arginine–apo-repressor complex.

There is no evidence for control by attenuation in this system.

6.10 Ribosomal proteins

There are 52 different proteins in the *E. coli* ribosome – 31 in the large subunit, 20 in the small subunit, and one common to both subunits. Many of these are encoded by genes linked together into several operons, some of which are very large (Table 6.6). A number of these operons also contain other genes encoding proteins involved in transcription or translation, including those for the subunits of RNA polymerase. A few of these genes seem to be unlinked.

There is good evidence that nearly all ribosomal proteins are produced in equal amounts, but it is not completely clear how this is achieved. Transcription of some of their operons is inhibited specifically by the binding of one of their protein products to the 5′-flanking region. These interactions occur with sequences which are homologous with part of the base sequence of the rRNA where the proteins are believed to be bound in the ribosome. This mechanism ensures that when there is an excess of a ribosomal protein which would be free and not actually incorporated into a ribosome, its synthesis and that of several other ribosomal proteins is shut off.

The operon encoding ribosomal proteins L10 and L7/12 and the β- and β′-subunits of RNA polymerase has been much studied. Under normal conditions of growth the ribosomal proteins are synthesised in much

Table 6.6. *The ribosomal protein operons*

rplK	rplA	rplJ	rplL	rpoB	rpoC					
L11	L1	L10	L7/12	pol	pol					
rpsL	rpsG	fusA	tufA							
S12	S7	EF-G	EF-Tu							
rpsJ	rplC	rplD	rplW	rplB	rpsS	rplV	rpsC	rplP	rpmC	rpsQ
S10	L3	L4	L23	L2	S19	L22	S3	L16	L29	S17
rplN	rplX	rplE	rpsN	rpsH	rplF	rplR	rpsE	rpmD	rplO	
L14	L24	L5	S14	S8	L6	L18	S5	L30	L15	
rpsP	orf	trmD	rplS							
S16			L19							
rpsM	rpsK	rpsD	rpoA	rplQ						
S13	S11	S4	pol	L17						

Gene symbols are given above the lines: the products they encode are shown below the lines. L and S refer to the proteins of the large and small ribosomal subunits respectively. EF are elongation factors required in protein synthesis (Chapter 2.5): pol are the RNA polymerase subunits. The proteins that are boxed can inhibit the transcription of all or most of the genes in the operon that encodes them (see text). The other ribosomal proteins are encoded either separately or in pairs. trmD encodes an enzyme which methylates G residues in some tRNAs. orf = open reading frame, coding for protein with no known function.

Table 6.7. *Some consequences of the stringent response*

Gene	Effect on transcription
rRNA	Inhibition
tRNA	Inhibition
Ribosomal proteins	Inhibition
EF-Tu and EF-G	Inhibition
lac operon	Stimulation
trp operon	Stimulation
galactokinase	Stimulation
ribulokinase	Stimulation
chloramphenicol-acetyl transferase	Stimulation

greater amounts than the polymerase subunits, although mRNA for all four proteins can be transcribed together, and the major promoter of this operon undoubtedly precedes the *rpl*J gene. There is a larger than usual non-translated spacer between the *rpl*L and the *rpo*B genes which contains several features of interest. First, it contains a potential rather weak terminator (attenuator) structure which could terminate transcription before the polymerase genes are reached. Secondly, there is a potential stem and loop structure, rather similar to that found in the intercistronic region between the 23S and 16S RNA genes (Chapter 5.3). This might be a site where RNase III could cleave the primary transcript, splitting off mRNAs for the polymerase which could be translated separately from those for the ribosomal proteins. Finally, there are two Pribnow boxes in this region which could function as independent promoters for the transcription of the polymerase genes, especially in circumstances where there is an increased production of the enzyme (e.g. after treatment with rifampicin).

In the operon which encodes ribosomal proteins S16 and L19, their genes are separated by the gene coding for the transmethylase enzyme which methylates guanine residues in some tRNAs and by an open reading frame encoding a protein of unknown function. The two ribosomal proteins are expressed at a level of about 8000 molecules per cell, while there are only about 80 transmethylase molecules per cell. It is not certain how these different levels of expression arise from the same operon. The transmethylase is over twice as large as either of the ribosomal proteins, so that translation of its mRNA would take much longer, and in addition it uses a number of codons that are not commonly used in more abundantly expressed *E. coli* genes. However, these two factors alone are unlikely to reduce the rate of synthesis of this protein by a factor of 100 compared to that of the ribosomal proteins. It is probable that there are other controls operating at the level of translation.

The *str* operon encodes two elongation factors used in translation in addition to two ribosomal proteins. EF-Tu is produced in about sevenfold excess over both EF-G and the ribosomal proteins. The *tuf*A gene has a promoter site about 50 bp upstream from the 3′-end of the *fus*A gene, so the former gene could be transcribed on its own. There is also a *tuf*B gene at a different site on the chromosome which encodes a protein functionally similar to, and very nearly identical with, EF-Tu. This gene contributes about a quarter or a half of the EF-Tu to the cell. This second gene is at the 3′-end of an operon which also encodes four tRNAs, and is the only operon known so far which encodes a mixture of stable RNAs and mRNAs. There is a strong promoter upstream from the first tRNA gene in this operon, but also a weaker one between the last tRNA gene and the *tuf*B gene, so there could be some transcription of the *tuf*B gene alone from this operon.

The diverse features encountered in these operons suggest that their individual functioning can be finely controlled in appropriate ways to serve particular physiological requirements.

6.11 The stringent response

Under conditions of amino acid starvation a co-ordinated series of events, known as the stringent response, takes place. These include cessation of the production of rRNA and tRNA and ribosomal proteins and inhibition of synthesis of carbohydrates, lipids and peptidoglycans, while simultaneously there is increased transcription of some of the genes for the synthesis of amino acids (Table 6.7). This response is mediated by a pair of unusual nucleotides, formerly known as Magic Spots I and II (they were originally identified on chromatograms). They are guanosine tetraphosphate – more strictly 3′-diphospho-,5′-diphospho-guanosine (ppGpp), and guanosine pentaphosphate (pppGpp) – bearing an extra phosphate on the 5′-position (Fig. 6.14).

Mutant strains of bacteria which do not exhibit the stringent response are known as relaxed strains. They have mutations in the *rel*A gene which encodes an enzyme mediating this response. This enzyme is a pyrophosphoryl transferase which is associated with the ribosomes and becomes active in the presence of uncharged tRNAs. The concentration of ppGpp increases dramatically more than tenfold when the enzyme is stimulated by the binding of uncharged tRNAs to the A site on ribosomes (Chapter 2). It is not known how ppGpp exerts its effects on the transcription of certain parts of the genome. A number of genes whose expression is inhibited in the stringent response have the conserved sequence CGCCNCC encompassing the site of initiation of transcription. This could possibly be a site at which ppGpp binds, or more likely, a binding site for a hypothetical control protein. Alternatively, ppGpp could bind to RNA polymerase. It has been suggested that this

Fig. 6.14. The formation of pppGpp and ppGpp, the magic spot nucleotides which are the effectors of the stringent response.

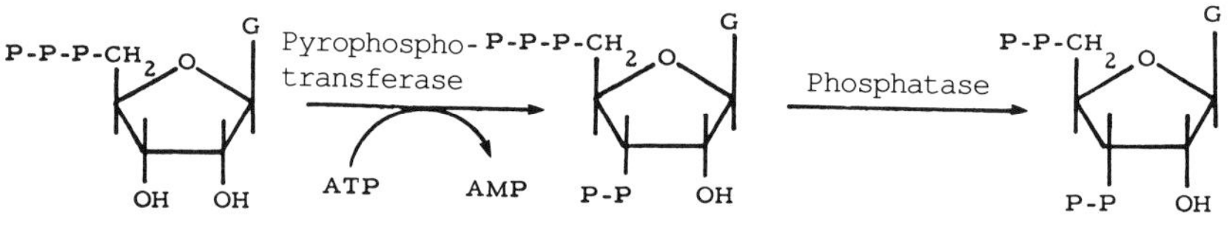

enzyme can assume different conformations which might recognise different promoters: if this were so binding of ppGpp could stabilise certain of these conformations, thus altering the efficiency with which it transcribes certain genes or operons.

7

Eukaryotic gene organisation and expression

7.1 DNA is in the nucleus in discrete linear chromosomes

Eukaryotic cells are distinguished from those of prokaryotes by the possession of a nucleus, separated from the rest of the cell by a membrane, and containing the cell's genetic material. (There is also a small amount of DNA in the mitochondria, discussed in Chapter 12.) The nuclear DNA is organised in linear chromosomes rather than in a single circular chromosome. The number of chromosomes varies considerably from one species to another (Table 1.1), and so does their size. There is little correlation between the total DNA content of a species and its evolutionary position. Many higher plants, amphibia and some fish have a much larger genome than humans and other mammals.

7.2 Nuclear DNA is associated with proteins

The nuclear DNA of eukaryotes is associated with a number of proteins which are found only in the nucleus. The best studied of these are a group called the *histones* which are among the most abundant cellular proteins. There are five major classes (Table 7.1) and, for proteins, they are comparatively small molecules containing an excess of positively charged amino acids (lysine and arginine) over the negatively charged ones (glutamate and aspartate). They are thus ideally suited for binding to negatively charged nucleic acids by multiple ionic interactions. Spermatozoa are exceptional in containing protamines, a group of small highly positively charged proteins, in place of histones. They presumably aid in packaging the DNA into the small volume of the head of the sperm.

As a group, histones have been very highly conserved during evolution. This is particularly marked in the case of H4 in which there are only two conservative amino acid differences between the proteins of cow and lily although animals and plants diverged from each other just over 10^9 years

Table 7.1. *The major classes of histones*

Symbol	Number of amino acid residues	Percentage of arginine residues	Percentage of lysine residues
H1	215	1.4	28.4
H2A	132	9.1	9.1
H2B	124	6.5	16.1
H3	135	13.3	8.9
H4	102	13.7	9.8

ago. The rate of divergence of other proteins having similar functions in different species (e.g. cytochrome c or haemoglobin) is faster by two orders of magnitude. This suggests that there are very stringent constraints on the structure of histones which are essential for their particular functions.

Many other proteins are associated with nuclear DNA. As they are present in low concentrations they are difficult to isolate and study, and little is known about most of them. One group that has been investigated to some extent are those that migrate rapidly on electrophoresis. These are known as the high mobility group. They are given arbitrary numbers, and two of them – HMG 14 and HMG 17 – may play roles in the control of transcription.

7.3 Histones associate in a regular fashion with DNA to form nucleosomes

If all the nuclear DNA of a single human cell were stretched out in its double helical form it would be about 230 mm long (on average about 5 mm per chromosome), but it is packed into a nucleus which is of the order of only 5×10^{-6} m in diameter. Thus the DNA must be packaged into a much more compact form. This is achieved in various ways which are not all clearly understood.

The fundamental unit of DNA structure in the nucleus is the *nucleosome*. Nucleosome preparations viewed in the electron microscope appear as a series of beads on a string. The beads are 10–11 nm in diameter, linked at regular intervals by a thin strand of DNA (Fig. 7.1). Mild digestion with micrococcal nuclease hydrolyses the DNA into units containing about 200 bp or multiples of this. Further digestion yields core particles which contain 146 bp. The DNA in these follows the course of one and threequarter turns of a left-handed superhelix wrapped round the outside of an octamer consisting of two molecules each of histones H2A,

H2B, H3 and H4 (Fig. 7.2). These protect the DNA from further digestion. The portions of DNA between these core particles are known as flexible linkers, and their length varies somewhat in different organisms and even between different tissues in the same organism. Histone H1 is bound to the outside of nucleosomes and probably plays a role in bringing individual nucleosomes together and stabilising them. This is facilitated because H1 shows co-operative binding to itself: that is to say, the binding of two or more molecules together enhances the ease of binding further molecules.

When DNA is associated with histone octamers to form nucleosomes it has to bend quite sharply. Trinucleotides of AAA/TTT are found preferentially where the minor groove of the DNA points inwards towards the octamer, while the dinucleotide GC/CG occurs frequently where the minor groove points outwards. This must place some constraints on the actual positioning of the nucleosomes on the DNA. Further compaction of the DNA is achieved by packing the nucleosomes together in regular structures with a higher degree of order. Electron microscopy shows that

Fig. 7.1. Chromatin fibres spilling out of rat thymus nuclei, showing nucleosomes between thin strands of DNA. (Reproduced from A. L. Olins & D. E. Olins, *Science* (1974), **183**, 330–2. © by the AAAS.)

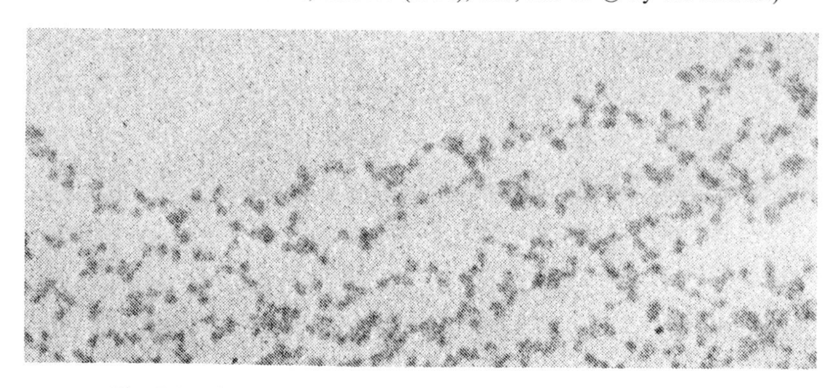

Fig. 7.2. The structure of a nucleosome. The circles and part circles represent the monomers of H2A, H2B, H3 and H4, while the tape winding round them represents the DNA double helix.

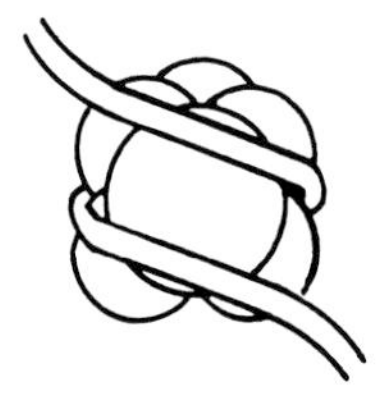

they form a fibre of 25–30 nm in diameter, in which they are probably packed into a solenoid-like structure (Fig. 7.3). These fibres, in turn, are formed into loops or domains of varying size containing from 35 to 85 kbp of DNA. Perhaps these loops are transcription units with some flanking sequences, since they are of the correct order of size (e.g. β-globin is transcribed from a region about 60 kbp long), but there is no direct evidence for this. It is believed that these loops are held in position by the enzyme topoisomerase II which acts as a scaffolding for attachment of the DNA.

Nucleosomes are probably in a dynamic state in which higher orders of structure can be temporarily disrupted by various processes, such as the removal of H1. This may be necessary for transcription to occur, but the evidence for this is largely circumstantial. Where there are discontinuities or gaps in the binding of the histones to form nucleosomes, digestion by an endonuclease can occur more easily giving rise to *DNase hypersensitive sites*. Their position is determined by gentle digestion with DNase followed by removal of protein and electrophoresis of the DNA fragments that are released. When these are compared to control digests of deproteinised DNA, fragments corresponding to protected sites will be missing (Fig. 3.4).

Examination of DNase hypersensitive sites has given interesting information about the state of DNA, particularly in the promoter and other significant regions of a number of genes. For example, the promoter region of the β-globin gene is resistant to digestion by DNase in cells where haemoglobin is not being synthesised, but readily digested and therefore free of nucleosomes in erythroid cells. The pattern of DNase hypersensitive sites around the lysozyme gene changes when its transcription is induced by steroid hormones in chick oviduct cells, while extra sites are found in macrophages where the gene is transcribed constitutively.

Fig. 7.3. Packing of nucleosomes. The larger circles represent the nucleosomes with the DNA double helix coiling round them. The smaller circles represent the H1 molecules bound to the flexible linkers of DNA.

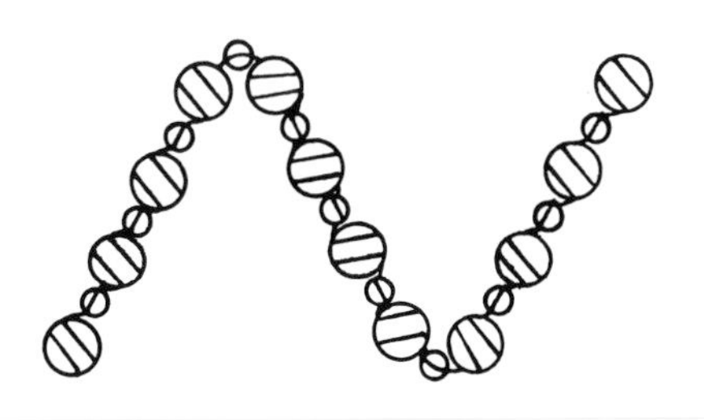

Table 7.2. *Histone gene clusters*

Species	Frequency of repetition	Repeat length (kbp)	Organisation
Sea urchin	300–600	5–7	H1 H4 H2B H3 H2A
Drosophila	100	25	H3 H4 H2A H2B H1
Xenopus	20–50	5.8	H3 H4 H2A H2B
			H3 H4 H2A H2B H1 H3 H4
Chicken	10	> 10	variable
Mouse	10–20	5.2	H4
Human	10–20	> 10	variable

Histones are particularly abundant proteins and they are synthesised at a specific stage in the cell cycle just before cell division takes place. Their genes were among the earliest to be studied. They usually occur in multiple copies in the genome (Table 7.2). In some invertebrates and in *Xenopus* they are clustered into tandemly repeated units. However, in other vertebrates and in yeast there do not seem to be regularly repeating units, though the genes are still clustered to some extent.

As a general rule, histone genes contain no introns, nor are their mRNAs polyadenylated, though there are a few exceptions to both generalisations.

The synthesis of new histones occurs in step with the replication of DNA in a tightly regulated manner. At the onset of replication, their genes are transcribed at an increased rate and the half-life of their mRNAs is increased, while when replication has finished the mRNAs are degraded more rapidly. Newly synthesised DNA seems to associate with 'hybrid' histone octamers that may contain both parental histones and newly synthesised ones.

Histones can be acetylated on lysyl residues and phosphorylated on seryl residues (Fig. 7.4). Both these processes decrease the positive charge

Fig. 7.4. Top: Acetylation of a lysyl residue in a protein. Bottom: Phosphorylation of a seryl residue in a protein.

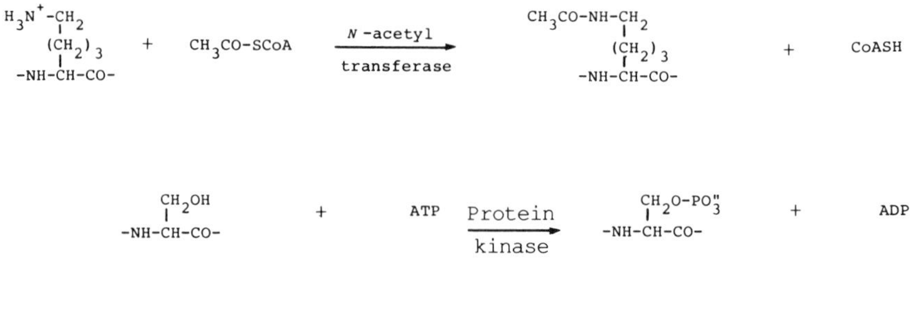

$$H_3N^+\text{-}CH_2\text{-}(CH_2)_3\text{-(-NH-CH-CO-)} + CH_3CO\text{-}SCoA \xrightarrow[\text{transferase}]{N\text{-acetyl}} CH_3CO\text{-}NH\text{-}CH_2\text{-}(CH_2)_3\text{-(-NH-CH-CO-)} + CoASH$$

$$CH_2OH\text{-(-NH-CH-CO-)} + ATP \xrightarrow[\text{kinase}]{\text{Protein}} CH_2O\text{-}PO_3^{=}\text{-(-NH-CH-CO-)} + ADP$$

on the molecules, so that they are likely to bind less tightly to DNA. Only certain residues, all situated near the N-terminus of the proteins, are subject to modification in these ways. The acetyl groups turn over rapidly. At any one time, about a tenth of the molecules of H2A are covalently linked to a small protein called ubiquitin, on account of its widespread distribution. The significance of this modification is uncertain, though other ubiquinated proteins are marked for degradation.

7.4 Eukaryotic replication

Replication of DNA is eukaryotic cells takes place, in outline, in a manner very similar to that in prokaryotes. However, there is a major difference due to the much greater size and the organisation of eukaryotic genomes. The measured rate of DNA replication in *Drosophila melanogaster* is about six times slower than in *E. coli*, so that it would take about 16 hours to replicate all the DNA in a *Drosophila* cell if replication proceeded from a single site. In fact, replication starts at a number of sites on each eukaryotic chromosome more or less simultaneously. Each unit where replication occurs is known as a *replicon*. Typically in the yeast genome which is not particularly large – only 14 Mbp – a replicon contains about 36 kbp and there are about 400 replicons in all. This means that replication can be completed in a few minutes.

In adult *Drosophila*, replicons are about 40 kbp long, but in embryos, where the rate of cell division is very high, replicons are only about 8 kbp long. It is likely that different replication origins can be activated in a developmentally regulated manner by specific proteins acting at different times.

Replication has been studied in animal cells infected with a virus such as SV40. This has a relatively simple genome that is circular, not linear. However, all the proteins involved in replication except one are those of the host. It is therefore believed to be a suitable model for the host's own replication processes. Replication begins at a definite origin (Fig. 7.5). In the first stage, a virally encoded protein, the T antigen, binds to repeated GAGGC sequences around the minimal origin and, with the aid of ATP, acts as a helicase to unwind the duplex DNA. Strand separation is aided by the A/T rich sequence in the centre of the origin. Two replication forks are exposed and kept open by single strand binding proteins(s).

The actual synthesis of the new strands is catalysed by two DNA polymerases. Polymerase α is a complex assembly of several subunits. The largest one contains the catalytic centre and there is a heterodimeric primase that catalyses the synthesis of RNA primers required to initiate the synthesis of the new DNA strand. This polymerase is not very processive and dissociates from the template after incorporation of about 1000 nucleotides into the new DNA strand. It is probably only involved in the synthesis of Okazaki fragments on the lagging strand. The more recently discovered polymerase δ is believed to catalyse the synthesis of the new DNA on the leading strand. It binds to an independently discovered protein – the proliferating cell nuclear antigen (PCNA), also known as cyclin – that confers a high degree of processivity on the polymerase. Polymerase δ appears to have no associated primase activity, and RNA polymerase II is probably used to make the primer for this strand. It is likely that the two DNA polymerases associate together to form a large asymmetric complex at each replication fork, reminiscent of the polymerase III in prokaryotic systems.

Unwinding of the duplex DNA at the replication fork places considerable torsional strain on the DNA ahead of the fork that is relieved by the action of a topoisomerase.

Some factors that stimulate transcription such as the protein Sp1 also stimulate replication. This is not too surprising since the two processes obviously have an overall similarity even though their detailed mechanisms must differ in important respects. There are at least six sites with the sequence GGGCGG where Sp1 could bind very near the replication origin.

Replication has also been studied extensively in yeast, whose genome contains a series of autonomously replicating sequences (ARS) that are used as replication origins for the 400 or so replicons found in this organism. An ARS consists of about 350 bp and contains five repeats of the sequence ${}^{A}_{T}\mathrm{AAA}{}^{T}_{C}\mathrm{ATAAA}{}^{A}_{T}$. Apart from its ability to be opened up fairly readily, little is known of its function. Replication is believed to

Fig. 7.5. The ori region of SV40. The upper lines with arrows show the direction of initiation and regions where primase starts replication. On the centre line, the smaller boxes to the left indicate the multiple T-antigen recognition sites 5′-GAGGC-3′, and the larger boxes to the right contain the Sp1 recognition sequence GGGCGG. There are also three T-antigen recognition sequences intermingled with these boxes that are not shown for clarity. Also shown are A- and T-rich sequences that may be sites where the T-antigen acts as a helicase to unwind the DNA duplex. The lowest line shows the minimal 65 bp sequence replication origin.

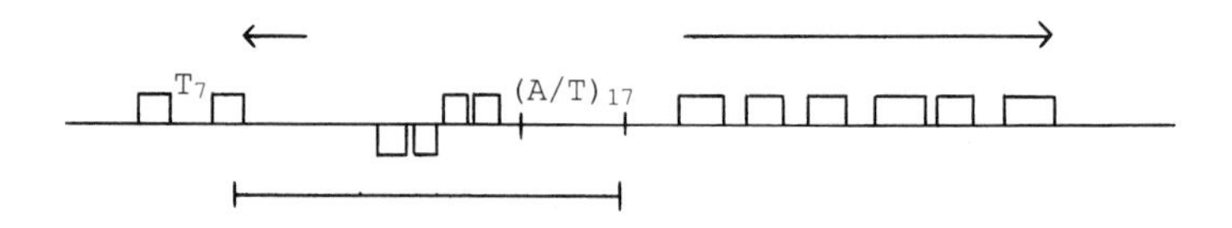

proceed in a manner analogous to that of SV40, but the nature of the analogue of the T-antigen is not known. The two polymerases in yeast are known as polymerases I and II rather than as polymerases α and δ.

DNA polymerases that have been purified from several eukaryotes seem much simpler than the holoenzyme from *E. coli.* They all have a similar structure with a large subunit containing the catalytic centre of the enzyme that is easily degraded during purification, two much smaller subunits with primase activity, and an intermediate sized subunit of unknown function. There are several sequences of amino acids in the catalytic subunit that are highly conserved between prokaryotes and eukaryotes. A number of other proteins such as a helicase, a proof-reading exonuclease, and RNase H are also required for DNA replication. Under some conditions these are found associated with the polymerases in much larger complexes.

There are two other eukaryotic DNA polymerases β and γ. The former is required for DNA repair while the latter is used in the replication of mitochondrial DNA.

7.5 Transcription

Transcription in eukaryotes is a good deal more complicated than in prokaryotes. Eukaryotic cells produce three different RNA polymerases for transcribing nuclear genes. These are all very large proteins with molecular weights in the range of 500 000 to 700 000, and each is made up of 10–15 distinct subunits. Some of these with molecular weights of around 200 000 are among the largest known single polypeptide chains. Each of the three different polymerases transcribes a particular set of nuclear genes. Polymerase I (Pol A) transcribes the precursors to ribosomal RNAs; polymerase III (Pol C) transcribes genes for tRNA precursors and some other small RNAs; Polymerase II (Pol B) transcribes genes into mRNAs coding for proteins. An RNA polymerase that transcribes mitochondrial genes is encoded by a nuclear gene. This enzyme seems to have a much simpler structure than those transcribing nuclear genes.

7.6 RNA Polymerase I transcribes ribosomal RNA genes

The 28S, 18S and 5·8S species of rRNA are encoded by large transcription units of variable sizes in different groups of organisms (Table 7.3). Nearly half the cell's transcriptional activity is devoted to transcribing the rRNA genes since 1×10^6–2×10^6 new ribosomes are needed for each new cell that is generated by cell division. The rRNA genes are located in multiple copies at the end of the short arms of five

Table 7.3. *RNA genes*

Species	Number of rRNA genes	Size of rRNA transcription unit, kbp	Number of 5S RNA genes	Number of tRNA genes
Yeast	140	8	150	360
Drosophila	180	8	150	750
Xenopus	600	8	500*	7000
			20000†	
Rat	160	13	830	6500
Human	150	13	2000	1300

* in somatic cells; † in oocytes
Some of these figures are the means of rather different estimates.

different chromosomes in humans. During interphase this DNA is looped out into the nucleolus for transcription by RNA polymerase I, which is confined to this organelle. This enzyme is readily distinguishable from the other RNA polymerases because it is insensitive to the fungal toxin amanitin.

In some amphibia and fish there is selective amplification of the rRNA genes during oogenesis so that there may be up to a million copies in a single egg. These amplified genes are found in extrachromosomal circular DNA.

RNA polymerase I recognises promoter sequences that are spread out to about 140 bp upstream from the site of initiation of transcription, though limited amounts of transcription can take place with artificially shortened promoters extending only about 40 bp 5′ to this site. There may be some enhancer sequences at various distances up to 2·2 kbp 5′ to the site of initiation of transcription.

RNA polymerase I has at least seven subunits of which three or more are phosphorylated by a tightly bound protein kinase. Phosphorylation increases the processivity of the enzyme, but has no effect on the rate of initiation of transcription. Two factors that stimulate the activity of this polymerase have been identified but not well characterised. One confers species specificity on the enzyme and binds to the promoter site.

Once the pre-RNA transcript has been made, it is very rapidly processed with about 600 nt cleaved off at each end to reduce the sedimentation coefficient from 47S to 45S, which was originally believed to be that of the primary transcript. The resultant molecule is then methylated and processed in an ordered sequence of specific hydrolytic cleavages to yield the mature 28S, 18S and 5·8S rRNAs (Fig. 2.3).

7.7 RNA polymerase III transcribes the genes of small RNAs

5S rRNA is transcribed from a different set of duplicated genes by RNA polymerase III. This enzyme, which is not associated with the nucleolus, also transcribes the genes for tRNAs and a few of the small RNAs.

Oocytes of *Xenopus* may contain up to 20000 copies of the 5S rRNA gene encoded separately from the rest of the rRNA genes (Table 7.3). They mostly occur in short, repeating sequences that also contain a pseudogene separated from the true genes by spacers of varying lengths. This very large complement of 5S rRNA genes is needed to keep pace with the synthesis of other kinds of rRNA that are transcribed from the amplified genes during the very early stages of growth immediately following fertilisation. Later in life, a different set of about 500 5S rRNA genes is used. These are scattered throughout the genome.

RNA polymerase III recognises promoter sequences that are inside the transcribed regions of the genes. Three such *internal control regions* (ICRs) have been delineated, and are referred to as boxes A, B and C (Fig. 7.6). The genes for 5S rRNA contain the boxes A and C to which essential transcription factors bind. Box A binds Transcription Factor IIIA (TF IIIA) rather weakly, and also binds TF IIIC, while box C binds TF IIIA strongly. The spacing between the two boxes is important.

tRNA genes contain boxes A and B and here the spacing seems to be much less important provided there are at least 21 bp between them. There are four copies of the gene for 7SL RNA (Chapter 2.4) in the human genome, though there are several hundred pseudogenes that are truncated copies. The four transcribed genes contain rather poor copies of the boxes A and B.

TF IIIA is only used in the transcription of 5S rRNA genes. It is found in high concentration in ribonucleoprotein particles in the cytosol of *Xenopus* oocytes bound to 5S RNA. It contains nine zinc fingers (Chapter 7.9) through which it binds in a 1:1 stoichiometry to 5S rRNA genes with a finger in the major groove every five nucleotides, covering much of the DNA between residues 50 and 97 downstream from the start of transcription.

Fig. 7.6. Top: 5S RNA gene, showing the boxes recognised by RNA polymerase III. Bottom: tRNA gene, showing consensus sequence of two boxes recognised by RNA polymerase III.

AGCTAAGCTGGG–N_{17}–TTGCATGGGAGACCGCCTG
Box A Box C

TRGCNNAGYGG——N_x——GGTTCGANTCC
Box A Box B

TF IIIC is probably a complex of at least two proteins, one of which binds strongly to box B, while the second one may contact box A. Less is known about TF IIIB, but it appears to bind after the other two transcription factors to make a very stable initiation complex to which polymerase III now binds. Until transcription is actually initiated this binding of polymerase III is not very firm. The enzyme can also dissociate fairly easily after transcribing a 5S or tRNA gene.

The snRNA U6 gene is unusual in that it can be transcribed by both polymerases II and III. *In vivo* it is probably mainly transcribed by the latter.

Transcription of all genes by polymerase III seems to terminate in a run of four or more consecutive T residues.

tRNA genes are present in multiple copies, often arranged in clusters in tandem, so the primary transcripts require processing to yield the individual tRNA molecules. Some of the precursors contain additional nucleotides in internal regions that must be removed by endonuclease action with subsequent re-formation of the appropriate phosphodiester link (Fig. 7.7).

7.8 RNA polymerase II transcribes genes coding for proteins and some snRNAs

All of the snRNA genes except that for U6 are transcribed by RNA polymerase II. The transcription of these genes is extremely efficient since over a million snRNA molecules are needed for each cell in each generation. Their promoters must be very strong, though surprisingly their sequences bear no resemblances to those for mRNAs. There is a moderately well-conserved element $\mathrm{GT}^{G}_{C}\mathrm{ACCGTGNGTRAAR}$ around 40–60 nt 5′ to the start of initiation of transcription, and a further element (YATGYARAT) with some of the properties of an enhancer about 150–200 nt further upstream.

There is also a signal with the consensus sequence $\mathrm{GTYYN_{0-3}A_3RRYAGA}$ 12–15 nt before the 3′-end of the genes.

Fig. 7.7. Portion of primary transcript of the gene for yeast $\mathrm{tRNA^{Phe}}$, showing the nucleotides that are excised during the formation of the mature tRNA.

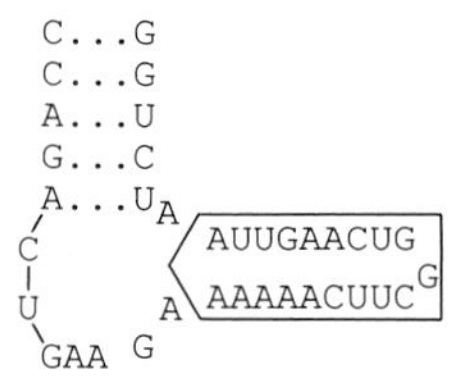

Table 7.4. *Sequences of TATA boxes in some eukaryotic genes*

G	184	20	4	5	4	5	64	124	251	220	188
A	101	24	554	63	532	435	459	297	186	103	150
T	89	499	41	527	58	157	59	146	69	92	71
C	225	58	2	6	7	4	19	34	95	183	189
Consensus	C/G	T	A	T	A	A	A	(A)	G/A	G/C	N

The figures show the frequency of occurrence of a particular base in a collection of 600 eukaryotic TATA boxes.

On the 5′-side (upstream) of the initiation codon of all genes coding for proteins a sequence of very variable length is transcribed into mRNA but never translated. Initiation of transcription is usually very precise, beginning at one particular nucleotide which is most frequently, though not invariably, an A residue. In some cases it may begin at several different nucleotides that are generally fairly close together. Upstream again from this initiation site (usually known as the cap site) is the promoter region. In the vast majority of genes, there is a highly conserved sequence homologous to the Pribnow box, termed the *Golderg-Hogness or TATA box* with the consensus sequence TATAAA (Table 7.4). This is usually situated about 30 nucleotides 5′- to the major site of initiation. Because base pairing is weaker between T:A pairs than between the C:G pairs that usually surround this site, the DNA double helix may be opened at the TATA box. This is believed to provide a 'bubble' to direct the RNA polymerase to start transcription at the specific initiation site. A number of genes lack a TATA box, and it is noteworthy that they generally show considerable heterogeneity in the site at which transcription starts. There are a number of protein *transcription factors*, among which TF-IID binds to the TATA box and, in conjunction with TFs-IIA, B and E, directs the binding of RNA polymerase II to this site before transcription actually starts. The largest sub-unit of this enzyme has multiple phosphorylation sites near its C-terminus, and there is evidence that the dephosphorylated form initiates transcription, while the phosphorylated form catalyses the elongation of mRNA.

Still further upstream there may be a number of other conserved nucleotide sequences (*upstream promoter elements* – UPEs) that play important roles in controlling the site and rate of transcription. These include the GCCAAT box and the sequence GGGCGG, frequently in multiple copies. Both may occur in the reverse orientation on the non-transcribed DNA strand. Distinct proteins that bind specifically to these sequences are widely distributed in nuclear extracts (Table 7.5). Several that have been purified are required for optimal *in vitro* transcription.

Table 7.5. *Transcription factors and their cognate DNA binding sites*

Factor	Cognate DNA binding sequence
AP1	TGA$^{G}_{C}$TCAG*
CREB†	TGACGTCA
CTF/NF1 } C/EBP }	GCCAAT
Sp1	GGGCGG
OTF1/NF-A1‡ } OTFII/NF-A2§ }	ATTTGCAT
Heat Shock Element	CTNGAANNTTCNAG
Metal Responsive Element	CTNTGCRCNCGGCCC
Interferons α and β	YAGTTTC$^{T}_{A}$YTTYYCC

Initials separated by a '/' are alternative designations
* Also a target site for a protein responsive to phorbol esters
† This factor is controlled by cAMP
‡ Found in many types of cell
§ Restricted to lymphoid tissue

7.9 DNA binding proteins

Many of the proteins that interact with DNA and control transcription have common structural features involved in binding to specific sites on the DNA. Generally, these reactions are highly specific depending on critical amino acid residues that determine this exquisite specificity.

Zinc fingers. Many DNA binding proteins contain one or more structures known as zinc fingers. These are sequences of about 9–13 amino acid residues where a zinc ion is co-ordinated to four cysteine residues at the base of the projection or finger (Fig. 7.8). Note that the spacing

Fig. 7.8. Zinc fingers. Left: the two adjacent zinc in steroid hormone receptors, with a consensus sequence shown in the one letter amino acid code. Right: the consensus sequence for a number of zinc fingers in several fungal proteins.

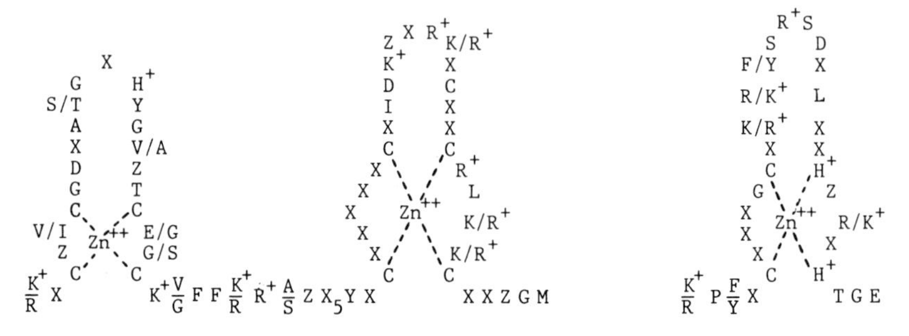

between the cysteine residues coordinating the zinc is variable. The fingers bind in the major groove of DNA, interacting with about five bases and may also interact with the negatively charged phosphoryl moieties in the DNA backbone through highly conserved cationic lysyl and arginyl residues. When there are several adjacent fingers they are believed to interdigitate with the DNA on the same side of the helix.

A family of fungal proteins containing only a single zinc finger controls the synthesis or catabolism of certain amino acids, carbohydrates and pyrimidines. The vertebrate receptors for steroid and other hormones all possess two closely adjacent zinc fingers. The specificity of binding of these proteins to particular sites on DNA is determined by certain residues either in these fingers or immediately adjacent to them.

In another class of zinc fingers, the zinc is co-ordinated to two cysteine residues and two histidine residues. The ubiquitous transcription factor Sp-1 that binds to the sequence GGGCGG contains three of these zinc fingers situated towards the C-terminal end of the molecule. The synthesis of truncated forms of this factor has shown that amino acid residues immediately adjacent to the zinc fingers are necessary for the activation of transcription, while some further away can modulate the strength of their binding to DNA. The *Xenopus* TF–IIIA that activates transcription by RNA polymerase III has nine zinc fingers. A number of genes important in the development and differentiation of tissues in *Drosophila* and vertebrates also contain multiple zinc fingers, and the *Xenopus* protein fin has no fewer than 37 of these structures comprising 90% of the protein!

Helix-turn-helix. Another structural motif commonly found in DNA-binding proteins in prokaryotes as well as in eukaryotes is the helix-turn-helix motif (Fig. 7.9). This consists of an α-helix containing eight amino acid residues, a β-turn of three residues and a second α-helix of nine residues. There are two of these proteins in the phage λ, known as the repressor protein and *cro*. The structures of their complexes with a 14-nucleotide oligomer to which they bind have been determined to high resolution by means of X-ray crystallography. Small differences in the amino acid sequence of these two proteins cause the conformation of the DNA to be altered slightly because of the interaction, such that in the complex with the repressor the DNA is straight, while in the complex with *cro* it is significantly bent. The second helix fits completely or partly into the major groove, while the first one sits above the DNA and makes contacts with the backbone of the DNA. In the different DNA binding proteins, some amino acid residues in this motif are highly conserved so that those with hydrophobic side chains are frequently on one face of the helix making van der Waals contacts with the bases. The residues whose side chains project outwards away from the DNA into the solvent are predominantly hydrophilic.

Leucine zipper. Some DNA binding proteins have lengths of α-helix containing leucine residues in every seventh position. Their side chains all project outwards on the same face of the helix and can make contact by interdigitating with similarly positioned leucine residues on another molecule of either the same protein to form a homodimer, or another protein, forming a heterodimer. There are generally four leucine residues on each helix in this interaction. Such a structure has been christened a 'leucine zipper', and gives rise to an extremely stable pairing that may be reinforced by other contacts between amino acid side chains. The yeast protein GCN4 forms a dimer in this fashion, and the products of the proto-oncogenes *jun* and *fos* are held together by this means. The c-jun protein, AP-1, can form homodimers with itself that bind rather

Fig. 7.9. Interaction of a helix–turn–helix motif with DNA. The two helices are shown as cylinders with the C-terminal one fitting snugly into the major groove of the DNA.

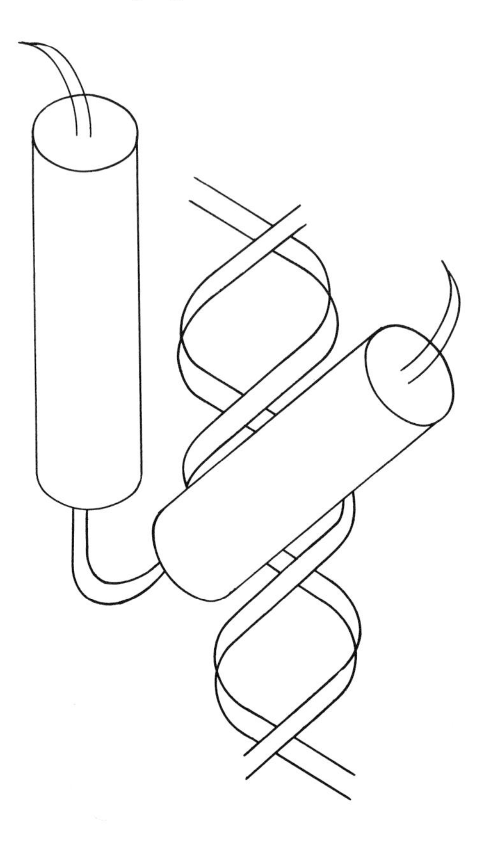

weakly to the sequence TGACTCAG, but the heterodimer with c-fos binds more strongly. c-fos can also bind to other members of a family of proteins similar to AP-1.

Just on the amino terminal side of these leucine zippers is a sequence of about 25 amino acids containing a high proportion of cationic side chains. This constitutes the DNA binding site of the protein. Site-directed mutagenesis has shown that mutations leading to replacement of the leucine residues in the zipper impair or abolish association of the polypeptide chains with each other, while mutations encoding the basic region abolish binding to DNA.

7.10 Transcriptional control

The protein AP-1, the product of the cellular counterpart, c-*jun*, of the oncogene v-*jun*, is a transcription factor that recognises and binds to the sequence TGA$^{G}_{C}$TCAG. There is a family of cellular *jun* genes, each encoding a different but homologous protein. It is not presently known whether these different proteins bind specifically to slightly different sequences on the DNA. In the cell, AP-1 exists largely as a heterodimer tightly bound through a leucine zipper to c-*fos* – the cellular counterpart of another oncogene, v-*fos*. The complex of AP-1 and c-*fos* binds more tightly to the appropriate DNA sequence than AP-1.

A number of other transcription factors and their cognate DNA binding sites have been identified (Table 7.5). It is not uncommon for several of these factors to bind in the promoter region of a particular gene. When this occurs the rate of transcription could be varied over a wide range in an exquisitively sensitive manner. Since several of the transcription factors are encoded by multi-gene families whose members are translated into slightly different proteins, this could further increase the sensitivity of control of transcription.

These UPEs and their cognate protein transcription factors are found in a variety of tissues. Other UPEs and DNA binding proteins that specifically enhance transcription in certain tissues only, or in response to certain substances or conditions are being characterised (Table 7.6).

The transcription of some genes is under the control of cAMP, which presumably acts by bringing about the phosphorylation of one or more proteins that bind to a responsive DNA sequence and activate transcription. Protein kinase C, that is activated physiologically by di-acylglycerols (DAGs) and by the tumour promoter phorbol myristyl acetate (PMA) – also known as tetradecanoyl phorbol acetate (TPA) – also causes the activation of certain genes. This again probably involves the phosphorylation of protein(s) that bind(s) to DNA in a sequence

Table 7.6. *Tissue specific and hormone response elements*

	Consensus DNA sequence of element	Number of sequences analysed
Tissue		
Liver	gGTTAaTrrTY$^{t}_{a}$aC	14
Pancreas	g$^{t}_{a}$cACctgt$^{g}_{c}$cTTTTCCctg	10
Hormone		
Glucocorticoids	gtacNtNTGTYCT	22
Oestrogens	gGtCAcNgTGaCC	7
Progestins	TGTTC$^{t}_{a}$Ct	7

Upper case letters indicate that the base occurs in > 80 % of the sequences.
Lower case letters indicate that the base occurs in < 80 % of the sequences.

specific manner. Interestingly, the DNA sequences that are responsive to these two similar stimuli differ only by a single base present in the cAMP responsive element, but absent from the protein kinase C one. While cAMP can direct the activation of both these regulatory DNA elements, the PMA responsive element is specifically activated only by this compound or DAG. The nucleotide context in which these sequences occur is important in determining whether or not a particular gene is controlled by cAMP. The transcription of several genes that have the sequence TGACGTCA fairly close upstream from their cap sites is not stimulated by cAMP. The reason for this is unknown.

Metallothioneine is a protein that binds very strongly to certain heavy metals, thereby reducing their toxicity. The transcription of its gene is activated when another protein specifically binds to one or more UPEs situated some way 5′- to the gene. Binding of this regulatory protein only occurs after it has itself bound a toxic metal such as cadmium.

When organisms or cells in culture are grown at a higher temperature than usual there are drastic alterations in the species of proteins that are synthesised. The synthesis of a special set of *heat shock proteins* (hsp) is switched on while the synthesis of some proteins that are normally made is greatly curtailed. These hsps are highly conserved over a wide range of organisms from insects to vertebrates and are even found in bacteria. They may also be produced in ischemia, inflammation and trauma, and in response to various toxic agents such as ethanol, free oxygen radicals and toxic metal ions. The genes coding for these proteins all have a highly conserved sequence (the heat shock element) in their promoter regions. In human cells a protein called heat shock factor is post-translationally modified when the cells are exposed to a temperature of 43 °C in such a way that it now binds to the heat shock element and presumably

Table 7.7. *Substances that bind to steroid hormone-like receptors*

Glucocorticoids	1, 25-dihydroxycalciferol
Mineralocorticoids	Tri-iodothyronine
Estrogens	Retinoids
Progestins	

stimulates the transcription of the neighbouring gene. This post-translational modification may be a phosphorylation, but other changes probably take place. It is not presently known whether this mechanism also functions in other organisms, but it seems a very plausible one.

Some of these hsps have been shown to be chaperonins – proteins that associate with newly synthesised polypeptides and are believed to provide a 'scaffold' that aids the correct folding of the protein into its tertiary structure. They may also be involved in the assembly of macromolecular complexes in mitochondria. Their increased production during heat shock could help to stabilise cellular proteins against thermal denaturation.

The actions of steroid hormones are mediated by members of a family of specific binding proteins. Different members of this family bind other relatively hydrophobic small molecules that act as hormones (Fig. 7.10 and Table 7.7). These proteins all consist of an N-terminal domain of very variable length, followed by about 60 amino acid residues containing two zinc fingers and a C-terminal domain that binds the hormone. When the protein is occupied by the hormone it is directed to bind through its zinc fingers to specific sites on DNA upstream from the genes whose transcription is being regulated. The N-terminal part of these binding proteins may play a part in directing the zinc finger domain to the specific binding site on the DNA. Unoccupied receptors are found in the cytosol complexed with hsp-90 (one of the heat shock proteins). When the appropriate hormone enters the cell it binds to the receptor and causes it to dissociate from hsp-90. The occupied receptor is now free to migrate to the nucleus and reacts with the appropriate DNA sequence. Mutation of some individual bases in the binding site on the DNA recognised by the glucocorticoid, progesterone and estrogen receptors alters their specificity in reacting to steroids of different classes in a predictable fashion.

Fig. 7.10 Schematic diagram of a steroid hormone receptor. Z is the zinc finger domain. H is the hormone binding domain.

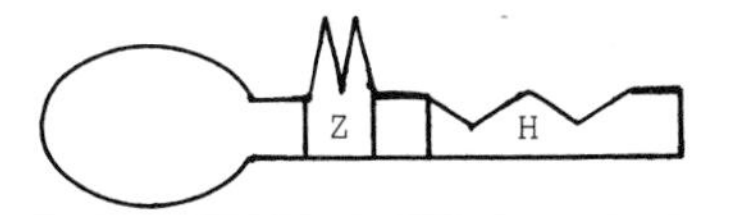

Binding of occupied steroid hormone receptors to DNA can also inhibit transcription in certain circumstances. For example, dexamethasone inhibits transcription of certain hormone genes in the pituitary. The precise molecular mechanisms by which the binding of these various proteins to DNA in promoter sequences exert their effects remain to be elucidated and are the focus of much active research at present.

7.11 Transcriptional control in yeast

The promoters of yeast genes are rather different from those of other eukaryotes. They generally have a TATA box, but this tends to be further away from the site of initiation of transcription – commonly between 40 and 100 nt upstream. Further upstream still, sometimes as much as 350 nt, there is always at least one *Upstream Activating Sequence* (UAS) that will specifically bind proteins.

A good example of these binding proteins is the product of the *GAL4* gene. This is a large protein normally present in very low concentration in the cell. It contains a single zinc finger that binds to the consensus sequence CGGA$^{G}_{C}$GACA occurring as a UAS in five genes (sometimes in multiple copies) that are required for the efficient utilisation of galactose. They are strongly induced when this sugar is present in the medium. In the absence of galactose, another protein, the product of the *GAL80* gene, binds to the GAL4 protein and prevents transcription from the promoters to which the latter binds. When galactose is present, a metabolite derived from it binds to, and causes the GAL80 protein to dissociate from the GAL4 protein so that it now binds to the UAS and stimulates transcription of the *GAL* genes.

GAL4 also binds to the UAS of the *GAL80* gene, increasing its rate of transcription, and thereby dampens the induction of the expression of the *GAL* genes induced by GAL4. Thus there is a continuing need for the presence of galactose for the production of the enzymes for its metabolism.

The transcription of genes encoding some of the enzymes required for the biosynthesis of the amino acids histidine, arginine, valine, leucine, isoleucine and tryptophan is also well regulated. The genes are not transcribed when the amino acids are present in adequate amounts, but, when they are lacking, the protein product of the *GCN4* gene binds to the sequence RRTGACTCATTTY upstream from these genes and activates their transcription. Amino acid starvation leads to an increase in the efficiency with which the mRNA of GCN4 is translated. The C-terminal region of the GCN4 protein is homologous to part of the vertebrate AP-1 transcription factor. In artificial constructs, these two homologous regions can be interchanged giving rise to proteins that still activate transcription when bound to the appropriate DNA sequences.

Both the GAL4 and GCN4 proteins contain sequences of about 60–100 amino acids that are predicted to form α-helices containing a high proportion of anionic amino acid residues, oriented to project out from the side of the helix. These regions bear a net negative charge and are responsible for the actual activation of transcription once the proteins have been bound to the DNA through another part of the molecule. Similar structures are found in some of the steroid hormone receptors mentioned previously. It has been suggested that these highly negatively charged domains ('negative noodles') may interact by H-bonds with the -OH groups in a heptapeptide repeated sequence that occurs near the C-terminus of the large subunit of all eukaryotic RNA polymerases II that have been examined. This has the consensus sequence $(YSPTSPS)_n$ where n may be as large as 52. It is not clear how this putative interaction between the negative noodle of the activating protein and the RNA polymerase II stimulates transcription.

7.12 Stability of mRNA

An important means of controlling the rate of production of proteins is by effects on the stability of mRNAs. Striking changes in the half-lives of some inducible mRNAs under various conditions are shown in Table 7.8. This is almost certainly due to changes in the rates of degradation, though other effects have not been ruled out. Degradation of mRNAs is believed to proceed from the 3′-end, but the nuclease(s) that perform this have not been well characterised, though some, at least, appear to be associated with polysomes.

Poly-A binding protein (PABP), a cytoplasmic protein that binds specifically to the poly-A tails of mRNAs (Chapter 7.15), stabilises poly-A containing mRNAs. It binds to a site containing approximately 27 A residues, and mRNAs with a smaller number of A residues are degraded so rapidly that they are generally undetectable. Antibodies to this protein or incubation of purified mRNAs in its absence greatly increase their rate of breakdown. The rate of dissociation of PABP from mRNAs could presumably control their stability.

The mRNAs for a number of cytokines and proto-oncogenes that normally have very short half-lives contain A- and U-rich sequences in their 3′-untranslated regions, where the motif AUUUA is particularly common and generally repeated several times (Fig. 7.11). The transposition of these sequences to the corresponding position in more stable mRNAs greatly increases their rate of degradation.

Some mRNAs are sequestered in translationally inactive forms as ribonucleoprotein particles in the cytoplasm where there are a number of proteins that can bind RNA fairly non-specifically.

Table 7.8. *The half lives of some mRNAs under various conditions*

mRNA	Tissue	Regulatory signal	Half life +Effector	Half life −Effector (hours)
Vitellogenin	Liver	Estrogen	500	16
Serum albumin	Liver	Estrogen	3	10
Ovalbumin	Oviduct	Estrogen + progesterone	24	5
Casein	Mammary gland	Estrogen	92	5
Insulin	Pancreas	Glucose	77	29
Histones	HeLa cells	DNA replication	4	0·25

7.13 The methylation of some cytosine residues may control transcription

It is relatively easy to study the degree of methylation of certain sites in DNA since the restriction enzyme Hpa II cleaves only the sequence CCGG when the Cs are unmethylated, while Msp I cleaves similar sequences irrespective of the state of methylation of the internal C. Thus, comparison of restriction maps produced with these two enzymes allows detection of methylated cytosine residues.

In the DNA of higher animals, about 70% of the cytosine residues in the dinucleotide sequence 5′–CpG–3′ are methylated in the 5-position, whereas most other cytosine residues are not. The methylation of cytosine does not affect its ability to base pair with guanine, so CpG base pairs are formed during replication whether or not the original cytosine residue is methylated. Many of these cytosine residues are maintained in a methylated state through repeated rounds of cell division. This is accomplished by an enzyme that specifically catalyses the methylation of newly incorporated cytosine residues in a 5′–CpG–3′ pair that is base paired to 3′–GpC–5′ which is already methylated (Fig. 7.12).

Fig. 7.11. Sequences in the 3′-untranslated regions of the human Tumour Necrosis Factor gene (upper) and mouse interleukin-3 gene (lower) that target them for rapid degradation. The characteristic AUUUA motifs are over- or under-lined.

UUAUUUAUUAUUUAUUUAUUAUUUAUUUAUUUA

UAUUUAUUUAUGUAUUUAUGUAUUUAUUUAUUUAUU

Table 7.9. *Genes whose transcription is inhibited by methylation*

Gene	Species and tissue	Effect
Thymidine kinase	Herpes simplex (virus)	Gene is hypomethylated when expressed. Expression of gene is inhibited by prior methylation
Adenosine-phosphoribosyl transferase	Mouse many tissues	Sequence 5′ to the coding portion of the gene is unmethylated in many mouse tissues where it is expressed
ε-globin	Human	Sites 5′ to the gene are unmethylated during expression in yolk sac and foetal liver. They become methylated at later times in foetal liver and in adult bone marrow when it is not expressed
γ-globins	Human	Sites 5′ to the genes are methylated in early foetal liver and in adult bone marrow when not expressed. They are unmethylated in later foetal liver when they are expressed
Vitellogenin	Cock liver	Methylation of the gene is decreased after oestrogen treatment which leads to its expression
β- and γ-casein	Rat liver and mammary gland	Genes are methylated in liver where they are not expressed, but hypomethylated during expression in the mammary gland

In vertebrate genomes the distribution of the dinucleotide 5′–GpC–3′ is random while 5′–CpG–3′ occurs very much less frequently than expected. However, CpG is present at the expected frequency in so-called islands of about 1–2 kbp that are commonly found near and overlapping the 5′-end of protein coding genes. They are known as *HTF islands*, since digestion with the restriction endonuclease Hpa II produces tiny fragments from these locations by cleavage at unmethylated C residues; elsewhere the CpG dinucleotides are generally methylated. There are about 30000 HTF islands in the haploid genome of most mammals, giving an average spacing of about 100 kbp, though there is great variation about this mean

Fig. 7.12. Maintenance methylation of newly incorporated cytosine residues after DNA replication

```
                                    5'—mC——G—3'                    5'—mC——G—3'
                                   ↗3'——G——C—5'                    3'——G—mC—5'
5'—mC——G—3'  Replication  /                      Maintenance
3'——G—mC—5' ——————————                          ——————————→
                           \                      Methylation
                            ↘5'——C——G—3'                       5'—mC——G—3'
                             3'——G—mC—5'                       3'——G—mC—5'
```

Table 7.10. *Genes where there is poor correlation between transcription and methylation*

Gene	Species and tissue	Effect
Immunoglobulin	Mouse plasmacytomas	Undermethylated in all expressed and in some unexpressed genes
Immunoglobulin	Mouse liver	Genes are methylated and unexpressed
δ-crystallin	Chick embryo	Demethylation follows induction of gene transcription in both lens and other embryonic tissues
rRNA	*Xenopus laevis* sperm	Gene expressed when non-transcribed spacer is fully methylated
rRNA	*Xenopus laevis* other tissues	Gene expressed when non-transcribed spacer is hypomethylated
Vitellogenin	*Xenopus laevis* liver	Gene expressed after stimulation by oestrogen with no change in level of methylation
Insulin	Rat pancreas insulinoma, and liver	No correlation between expression of the two insulin genes and their level of methylation

value. The deficit in CpG in the rest of the genome is attributed to the facile deamination of methylated C residues to T residues that have apparently become fixed since TpG occurs at a higher frequency than expected.

HTF islands have been found in all sequenced housekeeping genes that are highly expressed, but they also occur in a number of tissue specific genes. Some open reading frames encoding proteins of unknown function have been discovered very close to these islands.

In some invertebrates there is a completely different pattern of cytosine methylation, with long tracts containing unmethylated C residues, while others contain high levels of methylated CpG sites. There seems to be no association between the level of C methylation and the presence of genes. In other species, notably in *Drosophila* and in nematodes, virtually all the cytosine residues are unmethylated.

Whether HTF islands have any function is an open question, but it is likely that they are sites where proteins such as Sp1 and other transcription factors are bound. This binding probably limits the access of the enzyme catalysing the methylation at other CpG sites.

Methylation of the C residue in the recognition site for Sp1, GGGCGG, does not impair the binding of Sp1, or its effects in stimulating the transcription of a synthetic construct. However, methylation of critical C residues in the binding sites of other transcription factors such as CREB can inhibit both binding and transcription.

In a number of cases where genes are not being transcribed (e.g. in certain tissues or at certain stages in development), there are methylated cytosine residues in their 5′-flanking regions (Table 7.9). It is even possible to methylate cytosine residues 5′ to certain genes that are normally transcribed, introduce them into cells and show that they are no longer expressed. Methylation of these residues may provide a means for controlling gene expression. While this may be true in certain cases, there are a number of situations in which there is no apparent correlation between the degree of methylation and the level of transcription (Table 7.10). In no cases are there any clues to the factors in the cell determining the degree of methylation of cytosine residues that might be critical for gene expression. In fact, it is even possible that cytosine methylation or demethylation could be the effect of changes in the level of transcription rather than its cause.

A valuable tool in the study of methylation of DNA is the cytidine analogue 5-aza-cytidine (Fig. 7.13). It can be incorporated into DNA in place of cytidine, but it cannot be methylated and also inhibits the methylation of other cytosine residues.

When reticulocytes from adults are incubated with 5-aza-cytidine, cytosine residues in the 5′-region of the γ-globin genes are demethylated. These genes are normally only expressed in the foetus, but can now be transcribed in the adult cells.

7.14 Enhancers

The transcription of a number of genes is stimulated by the binding of specific protein factors to sequences known as *enhancers* that may be situated some way from the promoters of these genes. Enhancers have the remarkable property of operating over at least 3 kbp in either

Fig. 7.13. Cytidine (left) and 5-aza-cytidine (right), showing the 5-site of methylation in the former, and the impossibility of methylation at the 5-site in the latter.

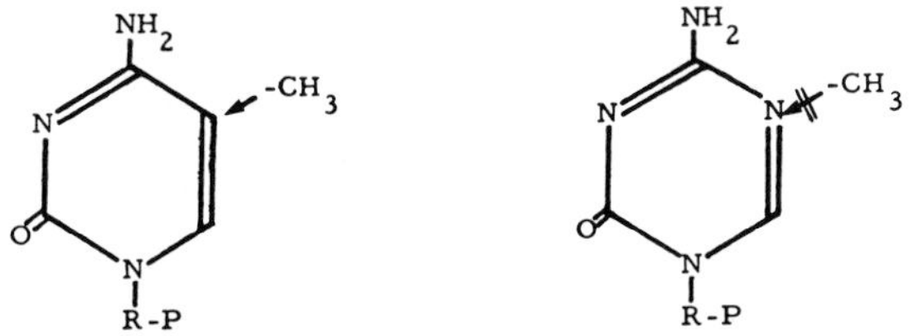

NH$_2$ -CH$_3$ N O N R-P

NH$_2$ N -CH$_3$ N O N R-P

direction from the start point of transcription and in either orientation (i.e. whether present in the $5' \rightarrow 3'$ or in the $3' \rightarrow 5'$ direction). They may allow RNA polymerase to bind to the DNA and move along the chromosome until it comes to a suitable promoter site where it can start transcription.

Homologous sequences with similar properties occur in the introns of immunoglobulin genes (Chapter 10.6). Although there is considerable divergence of sequence between these structures, a consensus sequence of GGTGTGG$^{AAA}_{TTT}$G has been derived.

These enhancers exhibit definite tissue or species specificity. Those associated with the immunoglobulin genes are only active in lymphoid tissues and several viral enhancers only function in the species in whose cells the virus grows best. This may be because tissue specific factors are needed for function. One such factor is NF-κB that interacts with the enhancer in the gene for the light chain of the immunoglobulins.

7.15 Termination

Almost all eukaryotic mRNAs that have been investigated, with the notable exception of those for histones, have a sequence of adenylyl residues (the poly-A tail) added post-transcriptionally in the nucleus by the action of the enzyme poly-A polymerase. Typically this tail contains about 50–200 A residues. They increase the stability of mRNAs by protecting them against degradation by ribonucleases, and they may have additional functions.

Although there are signals in the DNA to direct RNA polymerase to definite sites for initiation of transcription, the mechanism for ending transcription is not well understood. Downstream from the termination codon there is always a sequence of nucleotides that is transcribed and appears in the mRNA but is not translated. This is of very variable length that may, on some occasions, exceed 1000 nt. The sequence AATAAA is almost invariably found in all eukaryotic protein coding genes about 10–30 nt upstream of the site of poly-A addition. Natural or artificial mutation in this sequence interferes with proper 3′-end processing and may also cause increased transcription past the normal region of termination. Thalassaemias (Chapter 9.2) can arise by mutation of AATAAA to AATAAG or AACAAA in the poly-adenylation signal of the β-globin gene.

In some cases there are several AATAAA sequences in the portion of the gene 3′ to the termination codon. Poly-adenylation may take place specifically beyond one or other of these in particular circumstances.

Utilisation of these different sites may account for the observed heterogeneity in the size of mRNAs.

Important though this sequence undoubtedly is, it is not the sole determinant of the site of termination. Many genes have the sequence CAYTG a little downstream of the AATAAA and nearer the actual site of poly-A addition. Very commonly there is also a short sequence rich in GT and T residues just downstream of the site of polyadenylation. This is required for efficient termination since both deletion and mutation within it, as well as alteration in its distance from the normal polyadenylation site, significantly alter the efficiency of termination of transcription. Some proteins, in addition to poly-A polymerase, bind to this region, but the actual sequence(s) required for this have not been identified, nor is their function known.

There is also circumstantial evidence that some snRNAs, such as U 11, and their associated proteins may play a role in processing the 3′-ends of newly transcribed mRNAs.

When RNA polymerase II has incorporated labelled nucleotides for short periods of time, transcripts of a number of genes can be shown to terminate at very varied positions, ranging from 100–4000 nt beyond the poly-adenylation site. The termination sites for some genes occur over a distance of as much as 100 nt or more. It is not easy to detect these extended transcripts as they are very rapidly processed to generate the normal 3′-end of the mRNA. These results point to a random process in the termination of transcription but we do not understand what determines the site at which RNA polymerase II dissociates from the DNA template and ceases transcription.

Histone genes are small and compact (no introns), and their rate of transcription varies in a regular fashion through the cell cycle. It is greatest during S phase. Downstream from their termination codon is a small stem and loop structure, a spacer element whose actual sequence is not critical, and a conserved sequence CAAGAAACA (in sea urchins) and (less highly conserved) RAAAGAGCUG in vertebrates (Fig. 7.14). These conserved sequences can pair with the 3′-end of the small nuclear RNA U7 that is required for efficient termination of transcription of these particular genes.

The snRNAs that are transcribed by RNA polymerase III are not poly-adenylated and have a different termination signal, the 3′-box, that is situated immediately 3′ to a stem-loop structure present at the 3′-end of all snRNA genes. Surprisingly, it only functions when transcription is initiated from a snRNA promoter, since when this 3′-box is artificially inserted 3′ to the coding sequence of a gene encoding an mRNA, transcription is not terminated at this point.

7.16 Many mRNA molecules have a cap added after transcription

Eukaryotic mRNAs, unlike prokaryotic mRNAs, have a 7-methyl-guanyl cap at the 5′-end joined in a rather unusual way to the 5′-terminus of the transcribed mRNA (Fig. 7.15). Newly transcribed mRNA has a 5′-triphosphate (generally linked to adenosine) at the 5′-terminus. After this is enzymically hydrolysed to a 5′-diphosphate the capping enzyme, guanyl transferase, catalyses the addition of a guanyl radical, which is then methylated on position 7 by a specific methyl transferase (Fig. 7.15). Some viruses produce a single enzyme in which all three different catalytic functions reside in the same molecule. The 2′-positions of the ribose in the first two nucleotides at the 5′-end of the mRNA are frequently, but not invariably, methylated. Capping occurs rapidly on nascent mRNA chains while they are still being transcribed.

Capping of mRNA may serve two functions. First, it promotes the

Fig. 7.14. Transcription termination in vertebrate histone genes. The upper sequence is the consensus found at the 3′-end of histone genes. The lower sequence is U7 snRNA.

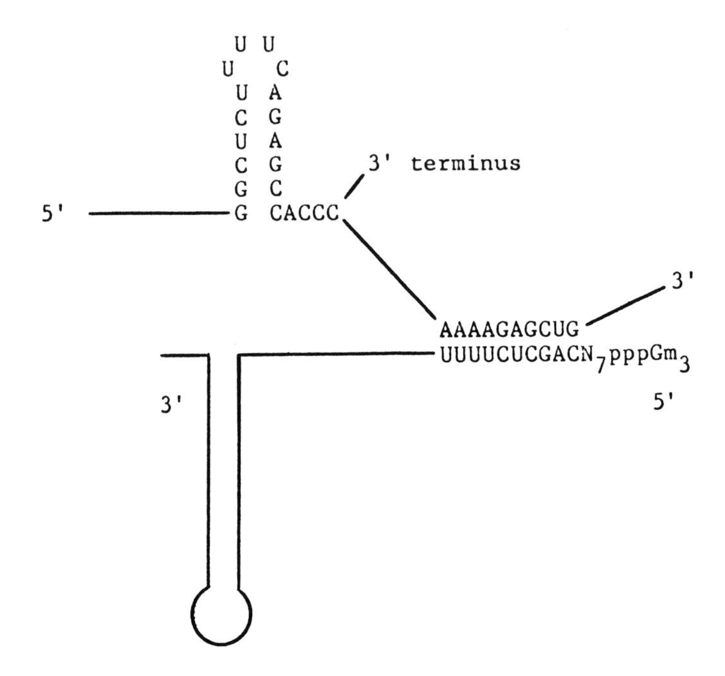

binding of the mRNA to the ribosomes. This is mediated by a cap-binding protein that binds to the cap through the positive charge bestowed on the guanyl residue by its methylation. Secondly, capping increases the stability of the mRNA, probably by protecting it against 5′-nucleotidase activity. Prokaryotic mRNA (lacking the cap) is very much more rapidly degraded than most species of eukaryotic mRNA.

7.17 The coding sequence of many genes is interrupted by non-coding sequences

In bacteria the mRNA is exactly co-linear with the genomic DNA from which it is transcribed. It came as a great surprise when it was found that this is not generally so in eukaryotes. Coding sequences are interrupted by non-coding sequences and the primary mRNA transcript

Fig. 7.15. The capping of eukaryotic mRNA.

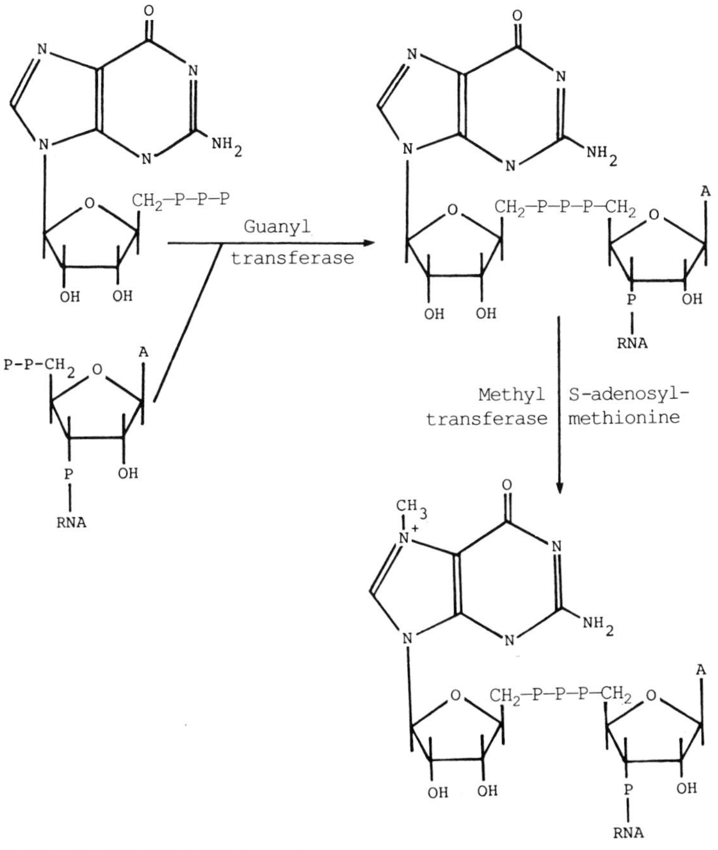

is processed by excision of the non-translated nucleotides. These non-coding sequences were originally called *intervening sequences* (IVS), but are more generally known as *introns*, and the coding sequences are called *exons*. Base sequences are fairly highly conserved at the 5′- and 3′-ends of introns, particularly in vertebrates (Table 7.11). However, there are fairly frequent exceptions to this, especially in fungi (other than yeasts) and insects.

Exons vary in length. Some are just a few nucleotides long, but in eukaryotes other than yeast few are longer than about 300 nt. In a random sequence of nucleotides (from which all protein coding sequences must ultimately have evolved) the chance of encountering a termination codon becomes very high in sequences of more than about 300 nucleotides. This may be the reason why introns are abundant. In fact, the highly conserved sequences at the beginning and end of introns (Table 7.11) both contain a termination codon, or could easily be derived from one by a single base mutation, so the most primitive proteins would probably have been short. Splicing of exons may have evolved because of the advantages that longer proteins would confer on the organism.

Introns are extremely rare in the genes of yeast, and the coding portions of the genes may be well over 1000 nt in length. Long exons of over 1000 nucleotides are found very rarely in vertebrates, but rather more frequently in plants, insects and fungi. Less than 5% of vertebrate genes (notably nearly all histone genes) contain no introns, while in the other groups of organisms about 20% of sequenced genes have no introns. Perhaps many of these intronless genes as well as the longest exons have evolved by loss of introns during evolutionary time. However, it is well known that transposons and other mobile genetic elements (Chapter 13.3) move around the genome and are inserted at various sites. It is likely that introns have been both gained and lost during the course of evolution. The varied exon/intron structure of some families of genes is described in Chapter 11.

Introns are much more variable in length than exons. The shortest known introns are just 31 nt long (in *Drosophila*), and the longest contain tens of thousands of nucleotides. There is presumably a lower size limit since they must contain the sequence information required for their removal.

The distribution of sizes of introns varies greatly between different groups of organisms. In *Drosophila* the majority of introns are less than 90 nt long, though there are a few very long ones, particularly those interrupting the 5′-non-coding parts of some genes. Introns are frequently found in the genes of fungi other than yeasts, where they tend to be short, with the vast majority containing less than 80 nt, and the longest are only

Table 7.11. *Sequences at the ends of vertebrate introns*

5′-end (5′ → 3′)

G	547	397	2334	2947	2	1183	348	2392	608	853
A	989	1765	294	0	1	1588	2074	219	488	721
T	350	426	204	3	2927	75	274	160	1354	551
C	1063	360	116	0	20	85	234	155	458	562
Consensus	C/A	A	G	↑ G	T	R	A	G	T	N

3′-end (5′ → 3′)

G	273	269	300	340	295	237	195	191	695	21	1	2952	1441	634
A	189	181	206	242	243	250	205	203	663	92	2952	1	792	694
T	1228	1268	1388	1318	1260	1248	1333	1483	712	628	0	0	307	971
C	805	784	890	1003	1116	1185	1215	1072	882	2211	0	0	412	574
Consensus	Y	Y	Y	Y	Y	Y	Y	Y	N	C/T	A	G ↑	G	N

The actual positions of the splice sites are indicated by vertical arrows.
The underlined sequences could be termination codons, or are only one base removed from them.
The numbers show the frequency of occurrence of a particular base in a collection of nearly 3000 sequenced vertebrate introns. The polypyrimidine tract in the lower sequence extends further in the 5′ direction.

about 400 nt long. Introns in genes of higher plants are generally somewhat longer but the majority are less than 150 nt long, and less than 5% are longer than 1000 nucleotides.

Vertebrate genes have introns of more varied lengths – only a third are less than 300 nt long, but, at the other extreme, about 15% are over 2000 nt long.

Among the yeast genes that do have introns, those for ribosomal proteins form the largest single group and the introns, usually 200–500 nt long, are found always near the 5′-end of the gene. Where they are present, there is only a single intron.

In some genes the total length of the introns may exceed that of the coding sequences by a factor of ten or more, so there seems to be much DNA that plays little or no functional role. Since a considerable amount of energy is used in synthesising intronic sequences that are subsequently excised and degraded and never translated, it is likely that their presence confers some biological advantage. For example, they could provide a pool of potentially useful DNA available for the production of new genes by mutation without the constraints imposed on sequences coding for specific proteins. Some introns contain potential regulatory sequences, such as enhancers, whose presence could provide further justification for the investment of energy in their synthesis.

Exons have undoubtedly moved around the genome (so-called exon shuffling), and some examples are given in Chapter 11.3. This may generate proteins with desirable properties. In some cases individual exons within a gene encode domains of protein structure that confer useful properties on the protein (e.g. specific binding to other molecules), but this is by no means universal.

Occasionally introns occur in the 5′- or 3′-flanking untranslated sequences of genes, so that there are whole exons that are not translated.

Introns can be detected and visualised by electron microscopy of heteroduplexes of mRNA and the genomic DNA from which it was transcribed (Fig. 3.13).

The rate of mutation in introns is higher than in exons since there are fewer constraints on the sequences of introns. This is apparent when the sequences of homologous introns and exons in the corresponding protein of different species are compared.

7.18 Introns are transcribed into RNA and then removed

Immediately after transcription, the introns in the transcript are rapidly removed by a process called *splicing*, in which several of the snRNPs (Chapter 2.3.4) are involved.

In the first stage, in the presence of Mg^{2+} and ATP, U1 RNA and associated proteins bind to the 5′-splice site (also known as the donor site) of the intron. Nucleotides adjacent to the 5′-terminus of this RNA can form base pairs with the conserved bases at the 5′-splice site. However, other factors must also be involved since some splice sites contain no bases that will base pair with the U1 RNA.

The U5 RNP binds at the 3′- or acceptor site of the intron, although base pairing is probably not involved since there is no region of the U5 RNA that is complementary to the conserved base sequence at this site. Binding of U2 RNP occurs around a particular A residue about 20 to 50 nt 5′ to the 3′-splice site where a reasonably well-conserved sequence can base pair with a sequence in the U2 RNA. Artificial mutation of this A residue inhibits splicing either strongly or completely. There is some doubt about the order in which the U2 and U5 RNPs actually bind. Finally U4 and U6 RNAs, bound to the same protein particle and to each other with considerable base-pairing, bind at or near the same site as the U2 RNP. This large complex (50S–60S) is called the *spliceosome*. It can be visualised in the electron microscope as a blob of stained material along a thin strand of pre-mRNA. Once the spliceosome is formed the U1 and U4 RNAs dissociate from it.

The actual excision of the intron starts with cleavage at the 5′-splice site and formation of a 2′-,5′-phosphodiester bond between the conserved A residue in the complex with U2, U5 and U6 and the G residue at the 5′-end of the intron. Then there is cleavage at the acceptor site, followed by ligation of the two exons and release of the excised intron as a lariat (Fig. 7.16). It is not clear whether these reactions are catalysed by the proteins or the RNA in the spliceosome complex.

In the few instances in which introns are removed from nuclear-encoded pre-mRNAs in yeast, the same general mechanisms seem to operate. However, the 5′-end of the intron has a much more highly conserved sequence (GUGAGUA), and the sequence round the branch point where the lariat is formed is completely conserved (UACUAAC), where the 3′-most A is the site of lariat formation. This latter sequence is also highly, though not absolutely, conserved in other fungi. The U20 RNA used in the splicing of introns in yeast is over 1000 nt long, while the homologous U2 in other eukaryotes has only rather less than 200 nt. However, the 100 nucleotides at the 5′-end of these snRNAs are highly homologous, and a yeast mutant with an internal deletion of 958 nt in the U20 RNA grows normally and splices the pre-mRNA for actin.

Since there is appreciable variation in the RNA sequences involved in the splicing of introns it is perhaps surprising that this process is always so exact, but this does seem to be the case. However, there are some circumstances in which the pre-mRNA transcript may be spliced in alternative ways, for example, in different tissues or at different stages of development, to direct the synthesis of distinct proteins, known as isoforms. There are many examples of this, but it is particularly common in transcripts of genes encoding some muscle proteins, such as

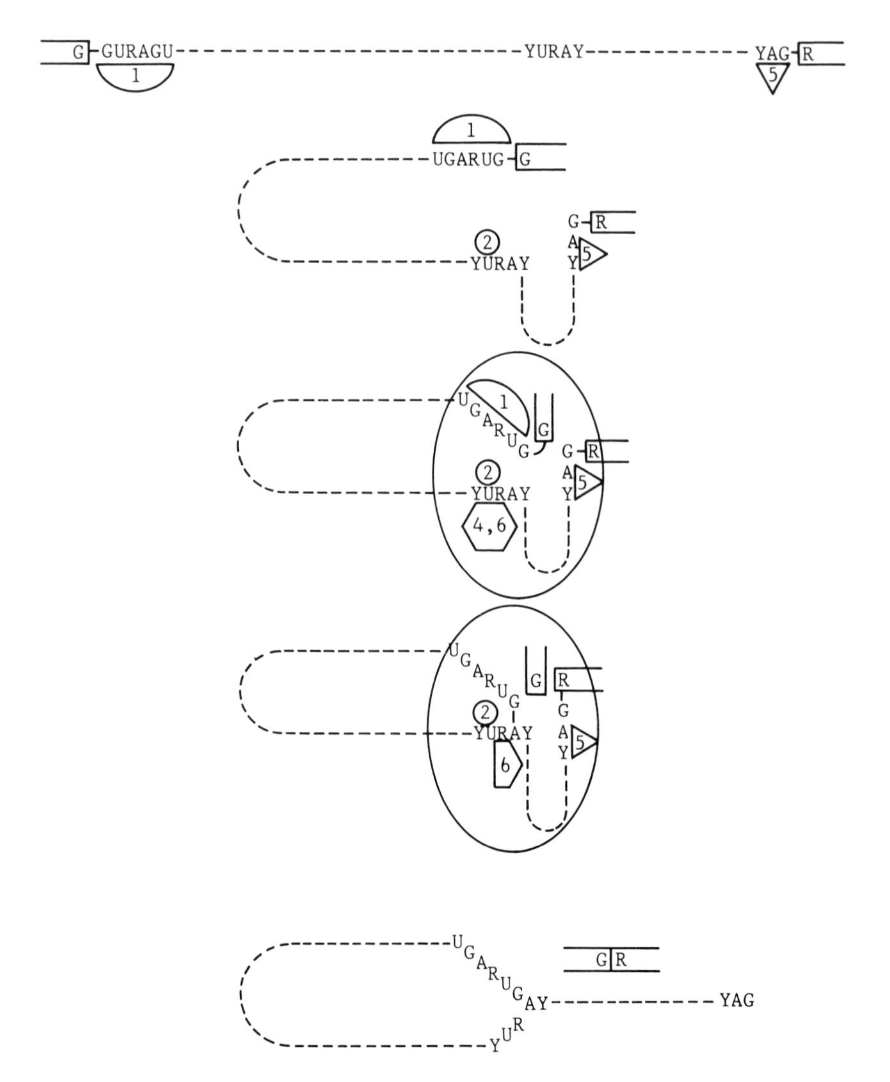

Fig. 7.16. Intermediates in the splicing process. The numbers refer to the snRNAs in their RNP complexes. The three base sequences are consensus sequences for splicing in vertebrates. The third and fourth lines show two stages in the spliceosome.

troponin-T, tropomyosin and the L and H chains of myosin. An example of two ways of splicing the transcript of the gene for the myosin alkali L-chain is shown in Fig. 7.17. In this case, the alternative splicing forces the two transcripts to use different promoters and cap sites, since two of the alternative exons are the first exons used for the two isoforms.

In *Drosophila*, alternatively spliced transcripts may give rise to non-functional mRNAs that contain premature termination codons. This mechanism generates two forms of transcript from the transposable P element. The functional form is found only in germ-line tissues and not in somatic tissues where the element is non-functional. Sex determination is also controlled by alternative splicing of transcripts of the transformer and double sex genes. One form of the product of the transformer gene controls the splicing of the double sex gene so that in females the gene product will turn off male differentiation genes, while an alternatively spliced form turns off female differentiation genes in males.

Most eukaryotic genes contain several introns, and spliceosomes may form on more than one intron at a time. Where it has been examined, splicing out of introns seems to proceed from the 5′-end of the gene.

Elimination of an intron in the transcript of the 26S ribosomal RNA gene of the ciliated *Tetrahymena* occurs in the absence of proteins. This involves a series of reactions where the O atom in the 3′-OH of a guanosine molecule makes a nucleophilic attack on the P atom of a phosphoryl group, temporarily liberating another 3′-OH group for a further

Fig. 7.17. Alternative splicing of the alkali L-chain myosin gene. The middle line shows the genomic structure of the gene. The upper form is found in embryonic and cardiac muscle. The lower form is found in adult skeletal muscle.

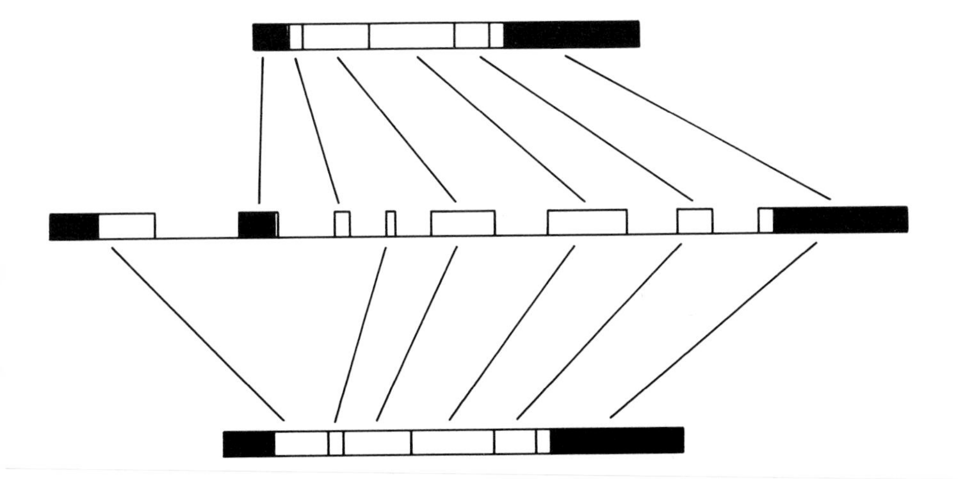

nucleophilic attack on another phosphoryl group. The intron is eliminated as a linear molecule that is subsequently cyclised (Fig. 7.18). A similar mechanism, in which proteins do not seem to take part, occurs in the splicing of the pre-mRNA of the yeast mitochondrial genome (Chapter 12.1). Since all these reactions involve transesterification there is no apparent need for an input of energy, and none appears to be needed in the case of the *Tetrahymena* and mitochondrial reactions, but ATP is hydrolysed during the processing of nuclear m-RNAs in the assembly of the spliceosome and in 5′- and 3′-cleavages.

An unusual form of splicing occurs in the nematode *Caenorhabditis*, and also in some ciliated protozoans. This is called trans-splicing since short lengths of RNA are transferred from one RNA molecule to the 5′-end of a pre-mRNA. Typically about 20 nt from SL RNA (a small RNA of only 100 nt) are used as the donor. It is estimated that this occurs in about 10% of *Caenorhabditis* mRNAs.

7.19 Post-translational modifications may be needed to produce functional proteins

Primary translation products may need further processing to acquire their full functional properties. An mRNA specifies only the amino acid sequence of the polypeptide chain that determines how the molecule will fold into its native secondary and tertiary structure. The folding probably occurs spontaneously so as to lead to the most thermodynamically favourable conformation.

Proteins that are destined for export from the cell are synthesised with an N-terminal extension of about 15–30 amino acid residues, generally

Fig. 7.18. Self-splicing in the *Tetrahymena* 26S ribosomal gene.

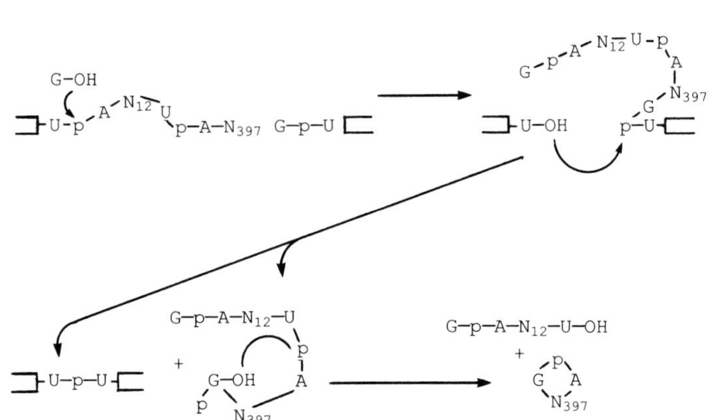

known as the *leader sequence* or *signal peptide*. In the middle of this sequence is a stretch of amino acids with hydrophobic side chains that has a high affinity for a *signal recognition particle* – a ribonucleoprotein containing the 7SL RNA (Chapter 2.3.4). This, in turn, binds to a *docking protein* on the endoplasmic reticulum that helps to draw the polypeptide chain, as it is synthesised on the ribosome, into the lumen of the endoplasmic reticulum where a signal peptidase removes the signal peptide by hydrolysis. After folding into the correct conformation, the protein passes through the Golgi apparatus before secretion from the cell.

Other kinds of proteolytic processing may occur, such as removal of an internal portion of the polypeptide to generate a protein with two (or more) polypeptide chains joined by disulphide links. This happens in the case of insulin. A common signal for the action of proteolytic-processing enzymes is a sequence of two adjacent positively charged amino acids (arginine and lysine). Examples of this are given in Chapter 11.

A wide variety of chemical groups can be added to proteins, particularly carbohydrates, since many proteins are glycoproteins. Glycosylation occurs in the Golgi apparatus immediately after removal of the leader sequence. This is often a very complex process in which many glycosyl residues may be added to a protein, some of which are later trimmed off. Certain sequences of amino acids are recognised by the enzymes involved in these reactions (Fig. 7.19). Because glycosylation requires a defined amino acid sequence there are considerable evolutionary constraints if it is a necessary feature of the structure of a protein.

Phosphorylation of serine, threonine and (less frequently) tyrosine residues is another common modification of proteins. This is often used to control the activity of enzymes, and again takes place within specific amino acid sequences containing the residue which is phosphorylated (Fig. 7.19). Phosphorylation is generally reversible, with a protein kinase

Fig. 7.19. Top: amino acid sequence recognised by enzymes capable of glycosylating proteins on an asparagine residue. X can be any amino acid. Bottom: amino acid sequence recognised by protein phosphokinases. X, Y, Z can be any amino acid residue.

Asn X Ser/Thr

Lys Arg X Y Ser Z
or
Arg Arg X Ser Y

used to phosphorylate the protein and a protein phosphatase to remove the phosphate group.

Another modification is acylation (generally with an acetyl radical) of amino groups – either the N-terminal ones, or those on lysine side chains (Fig. 7.4). This reaction changes the electric charge on a protein, and has already been discussed in connection with the histones (Chapter 7.3).

8

Oncogenes

8.1. Retroviruses

Much interest has been aroused in the family of retroviruses because they can be oncogenic – that is to say they may cause cancer. Their genomes consist of fairly short (about 5–9 kbp) strands of RNA containing a very limited number of genes. They are encapsulated in a protein coat that is encoded by two of these genes. The *gag* gene directs the synthesis of its core protein, and the *env* gene codes for the glycoprotein that occurs as a spike on the surface of the envelope. They may also carry *oncogenes* (genes that cause cancer) in their genome (Fig. 8.1).

Replication of these viruses is initiated by a reverse transcriptase that is encoded by their *pol* gene. When cells become infected with such a virus the host's RNA polymerase transcribes this gene, producing an mRNA that directs the synthesis of reverse transcriptase on the host's own ribosomes. This enzyme than transcribes the viral RNA into a single-stranded DNA molecule which directs the synthesis of a complementary strand of DNA making use of the host's DNA polymerase. This double-stranded DNA is then integrated into the host's genome as a provirus. Integration can occur at many sites with duplication of 4–6 bp of the host DNA sequence at each end of the insertion. The provirus can now be replicated or transcribed in the usual way, resulting in either the spread of the viral sequence among the host's cells or the production of new virus particles.

The 5′-end of retrovirus RNA contains a cap structure (Chapter 7.16) followed by a short sequence of up to 80 bases (R), that is also found at the 3′-end of the genome. This is succeeded by a sequence of about 100 bases (U5), preceding the coding portion of the genome. A longer sequence of up to 1000 bases (U3) occurs at the 3′-end of the coding sequences, followed by the R sequence and a poly-A tail. When the RNA

Table 8.1. *Some oncogenes found in retroviruses*

Oncogene symbol	Name of virus
Tyrosine protein kinases	
v-*src*	Rous sarcoma virus
v-*abl*	Abelson leukemia virus
Serine protein kinase	
v-*mos*	Moloney murine sarcoma
G-related proteins	
v-*Ha-ras*	Harvey murine sarcoma
v-*Ki-ras*	Kirsten murine sarcoma
Nuclear localised proteins	
v-*myc*	Avian myelocytomatosis
v-*jun*	Avian sarcoma 17
v-*fos*	FBJ murine sarcoma
v-*erb A*	Avian erythroblastosis
Growth factors or their receptors	
v-*sis*	Simian sarcoma
v-*erb B*	Avian erythroblastosis
v-*fms*	McDonald feline sarcoma

is transcribed into DNA, these end structures are duplicated and rearranged so that each end now has a long terminal repeat (LTR) consisting of U3-R-U5 (Fig. 8.1). This structure is very reminiscent of that found in transposons. The U3 unit in the integrated proviral DNA contains a sequence related to the TATA box, and also the polyadenylation signal, AATAAA, that are presumed to be important for transcription and for poly-A addition to the transcribed RNA respectively.

Oncogenic viruses (Table 8.1), in addition to causing cancer when injected into suitable hosts, often change the growth characteristics of certain cell types in culture. This latter action is called transformation, and, since it can be readily detected, it is widely used in studies with these viruses.

Fig. 8.1. Top: Retrovirus organisation, *gag*, *pol* and *env* are the viral genes (see text). Bottom: Retroviruses after reverse transcription and integration into the host genome. Host genes can be incorporated anywhere into this structure.

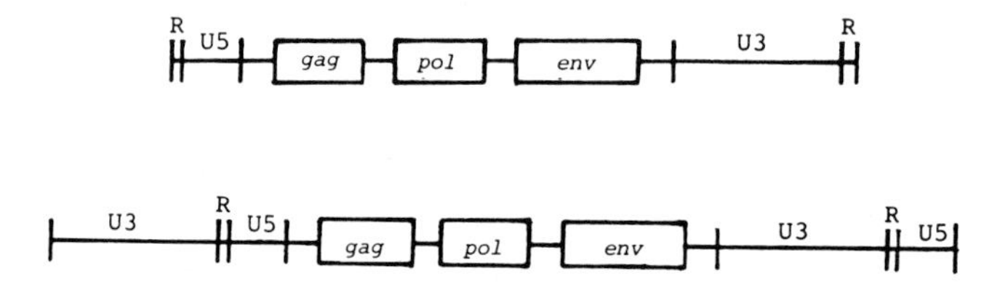

Oncogenic viruses are believed to have arisen when a retrovirus has integrated very near to certain genes that have critical functions for cell growth. By a rare event such a gene can be incorporated into the proviral DNA and thence into the viral RNA genome. It is suggested that this gene is transcribed more efficiently when it is situated near a viral promoter than when it is under the control of the cell's own promoter. Thus, the gene's product will be synthesised in larger quantities than usual leading to unrestricted growth (cancer). A number of genes that are normal components of animal genomes are very similar to viral oncogenes. These are known as proto-oncogenes, and are given the same symbol as the viral gene, except that they are preceded by the prefix c-. Since the viral oncogenes lack introns it is likely that they have arisen by reverse transcription of cellular mRNA. They have also undergone a small number of point mutations. Because of the extensive homologies between the viral and cellular genes, it is relatively easy to isolate the latter by hybridisation, and some of their sequences have been determined. They contain open reading frames that are capable of encoding proteins.

8.2 Oncogenes

The proteins encoded by the proto-oncogenes fall into four groups – protein kinases; guanine nucleotide binding proteins that are believed to mediate the action of some hormones; proteins localised in the nucleus that may play roles in controlling transcription; and proteins that are either growth factors or their receptors.

The protein kinase proto-oncogenes mostly act by phosphorylating tyrosine residues on proteins – a comparatively rare reaction that is probably involved in the control of cell growth, since the receptors for several hormones or growth factors possess this capability. The proto-oncogenes c-*erbB* and c-*fms* encode receptors for epidermal growth factor and macrophage-colony stimulating factor respectively and their action is mediated by activation of a tyrosine protein kinase domain in the receptors. In the oncogenes v-*erbB* and v-*fms* this activity is believed to be permanently activated. A smaller number of the proto-oncogenes (e.g. c-*mos*) possess serine/threonine protein kinase activity.

When mutations in protein kinase genes result in either enhanced activity of their gene product or loss of normal transcriptional regulation, it is not difficult to envisage that uncontrolled growth could ensue.

The *ras* family of oncogenes is homologous to the family of G proteins that mediate the interaction between some occupied hormone receptors and either adenylate cyclase or phospholipase C (the effector enzymes responsible for producing the second messengers cAMP and inositol

Table 8.2. *Mutations in* c-Ha-ras *found in* v-Ha-ras *and various human carcinomas*

Carcinoma or oncogene	Codon 12	Amino acid encoded
c-*Ha-ras*	GGT	Glycine
Bladder carcinoma	GTC	Valine
Colon carcinoma	GTT	Valine
Lung carcinoma	TGT	Cysteine
v-*Ha-ras*	AGA	Arginine
v-*Ki-ras*	AGT	Serine
Mammary carcinoma*	GAA	Glutamate

* Induced in the rat by treatment with nitroso-methylurea.

trisphosphate respectively). When these proteins bind GTP they activate the effector enzymes but this effect ceases when the G protein hydrolyses GTP to GDP. The oncogene products bind GTP, but hydrolyse it much more slowly so the systems are switched on for relatively long periods.

In cells cultured from a number of different carcinomas the normal human homologue of the rat proto-oncogene c-*Ha-ras* has suffered point mutations in a specific codon that is also mutated in the related viral oncogenes v-*Ha-ras* and v-*Ki-ras* (Table 8.2). The normal human gene encodes glycine at this position, and mutations to codons for serine, cysteine, valine or arginine have taken place in four different carcinoma cell lines and in the viral oncogene. When all possible amino acid substitutions were made at this position by site-directed mutagenesis, they all produced oncogenic proteins capable of transforming cells *in vitro*, except when proline was placed there. Proline and the naturally occurring glycine both have very strong helix-breaking potential so it is probable that the conformation of the protein is substantially altered by the substitution of any amino acid except proline for glycine at this position.

Substitution of valine for glycine at position 12 in the proto-oncogene product causes a bulge in the pocket where GTP is normally bound. This probably accounts for the less efficient hydrolysis of GTP by the oncogene product.

The proto-oncogene products jun and fos are found in the nucleus where they play important roles in controlling transcription (Chapter 7.10), and it is believed that other proto-oncogene products localised there may have similar functions. When cells are infected with the oncogenic avian sarcoma virus it appears that the normal controls governing the rate of transcription of the jun gene are abrogated so the protein product is produced in larger amounts than normal. In the case of the FBJ murine osteosarcoma virus, sequences in the untranslated 3′ portion of the gene are absent. In the proto-oncogene these contain the sequence ATTTA that is believed to lead to rapid degradation of the mRNA once it has been formed (Chap. 7.12). Thus the oncogene mRNA is not rapidly degraded, leading to accumulation of the fos protein in the cell.

The v-*erb A* oncogene is derived from the normal gene encoding the receptor for the thyroid hormones (Chapter 7.10). Its product encodes a protein with an intact DNA binding domain, but it can no longer bind these hormones. When present in a cell, it binds to the receptor binding site on DNA and acts as a repressor of genes whose transcription is normally stimulated by thyroid hormones. It has been suggested that the v-*erb A* product blocks the normal differentiation of erythroblasts that is thyroid-dependent, leading to unrestricted and undifferentiated growth.

The oncogene *v-sis* encodes a mutated form of the platelet-derived growth factor that is believed to keep its receptor in a permanently activated state.

Some carcinogens seem to bring about mutations that convert proto-oncogenes into oncogenes in a fairly reproducible fashion. For example, the *Ha-ras* oncogene is activated in rats by *N*-nitroso-*N*-methylurea causing mammary carcinomas and, by mutation at a different site, in mice by dimethylbenz(a)anthracene causing skin carcinomas.

Retinoblastoma is a cancer occurring inside the eye in infants and there is a strong hereditary component in its incidence. It is probably caused by lack of the product of the retinoblastoma gene on chromosome 13 due to deletions at this locus. The gene encodes a nuclear phosphoprotein, but its function is unknown, though it has strong anti-oncogenic potential and may normally limit the growth and differentiation of retinal cells. Levels of the product of the *n-myc* gene are generally elevated in retinoblastoma patients but the reason for this is unknown, though this could contribute to their symptoms. The oncogenic adenovirus E1A produces a protein that binds to the retinoblastoma gene product. This probably inhibits its normal activity and so leads to cancer. It is believed that there may be a number of other anti-oncogenes, the loss of whose products causes cancer. They presumably limit the growth of cells that would otherwise proceed in an uncontrolled fashion.

These fascinating discoveries suggest that cancer may arise both by over- or under-production of some specific cellular proteins that are normally required for tight control of growth, and also by subtle alterations in their structure and functions.

Table 8.3. *Chromosomal translocations which give rise to cancers*

Nature of disease	Chromosomal translocation	Proto-oncogene near to position of translocation
Leukaemia	t(9:22)	c-*abl*
Leukaemia	t(8:21)	c-*mos*
Burkitt's lymphoma	t(8:14)	c-*myc*
Burkitt's lymphoma	t(8:22)	c-*myc*

8.3 Chromosomal alterations in cancer

It is becoming increasingly clear that neoplasias (cancerous conditions) are very frequently associated with various chromosomal abnormalities, particularly deletions of parts of chromosomes, and translocations in which portions of non-homologous chromosomes are exchanged. Deletions are generally, though not exclusively, associated with solid tumours, while translocations are more usually found in leukaemias and lymphomas where there are cancerous proliferations of various kinds of leukocytes. These translocations are sometimes near the known sites of proto-oncogenes (Table 8.3).

A very early discovery in this field was the so-called Philadelphia chromosome, found in many patients with certain types of leukemia. This results from a translocation between chromosomes 9 and 22 (symbolised

Fig. 8.2 Structure of human c-*myc*, and part of the human Ig μ gene. The other structures are of various translocations between these two loci that have been isolated from cells of patients with Burkitt's lymphoma. The c-*myc* exons (E1–3) are shown below the line: the Ig exons above the line.

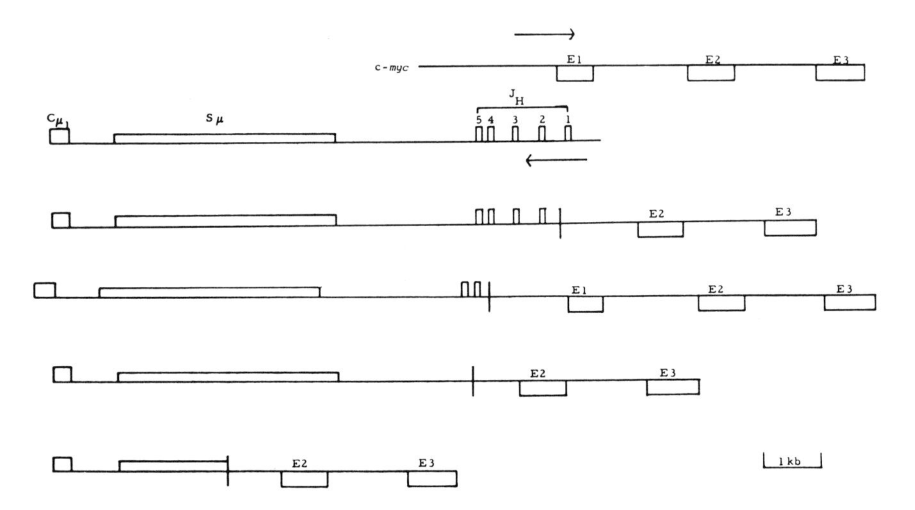

t(9:22)). The break point on the former is near the proto-oncogene c-*abl*, and that on chromosome 22 is close to the gene for the immunoglobulin λ chain.

One of the most studied conditions is Burkitt's lymphoma in which a part of the end of chromosome 8 has been exchanged with a sequence near the end of chromosome 14 (Fig. 8.2). The segment from chromosome 8 contains the proto-oncogene c-*myc*, and it is translocated into the region of chromosome 14 where the cluster of H-chain immunoglobulin genes is found (Chapter 10). These translocations are not all found at exactly the same point in different patients. In a few cases of Burkitt's lymphoma there is translocation of the same segment of chromosome 8 to sites on chromosomes 2 or 22 where the L-chains of the immunoglobulins are encoded. Analogous translocations of the chromosomal sites for c-*myc* (chromosome 15) and that for the H chain genes (chromosome 12) are found in some mouse plasmacytomas (see Chapter 10). Increased amounts of the RNA transcripts of the c-*myc* gene are frequently, though not invariably, found in cells where these translocations have occurred. The significance of this is not apparent, since we do not know the normal function of the protein encoded by the c-*myc* gene. If, as seems possible, the c-*myc* gene product is concerned with cell growth in some way, over-production could obviously cause increased cell proliferation and a cancerous condition. The detachment of the c-*myc* gene from its usual chromosomal position could result in loss of some control elements so that it is expressed in greater amounts than usual in its new chromosomal environment. The chromosomal deletions observed in other types of cancer could also lead to loss of control over the normal expression of genes concerned with normal growth.

Several other sites where translocation occurs in neoplasias are near oncogenes, so that similar mechanisms may operate in these cases. A few neoplasias are associated with the condition of trisomy, in which there are three copies of a particular chromosome instead of the usual two. This could result in over-expression of a gene giving rise to increased concentrations of a protein factor promoting uncontrolled cellular proliferation. In the mouse, trisomy of chromosome 15 which bears the c-*myc* gene, is frequently found in leukaemia.

Cytogenetic studies have identified so-called fragile sites at which chromosomes are particularly likely to become broken, and some of these are the sites at which translocations frequently take place.

9

Haemoglobin

9.1 Genes for globins are found in two clusters

Haemoglobin consists of four polypeptide chains of two different though similar types, which are folded round each other in an orderly and compact fashion. The two types of chain, designated α and β, are present in equal amounts. They show a very substantial degree of homology – in humans 43% of the residues are identical. During manufacture of haemoglobin these polypeptides (known as globins) are synthesised first and then each one binds a molecule of haem very firmly.

Several different β-like globins are synthesised at different stages of human life (Table 9.1). In the early embryo when haemoglobin is synthesised in the yolk sac, the ε chain is made. Later, synthesis switches to the foetal liver and two forms of the γ chain are made. In one of these an alanine residue is found at a position where the other contains glycine, and they are therefore known as $^{A}\gamma$ and $^{G}\gamma$ respectively. Finally, just before birth, β-chain synthesis commences, along with that of a very small amount of the nearly identical δ chain which differs from the β chain in only 10 residues.

The δ gene is transcribed at a much lower efficiency than the β gene so that δ chain mRNA is produced in much smaller amounts than β chain mRNA. There are sequence differences in the 5′-flanking regions of the δ and β genes which probably account for this (Fig. 9.1), but it is not yet possible to pinpoint the particularly crucial differences.

There is also a developmentally regulated family of α globins. In the yolk sac embryo ζ chain is synthesised, while both the foetus and the adult produce the α chain. Switching from ζ to α and from ε to γ is not precisely co-ordinated since small amounts of the tetramers $\alpha_2\varepsilon_2$ and $\zeta_2\gamma_2$ are found during this transition.

These diverse forms of similar molecules have arisen during the course

Table 9.1. *Haemoglobins formed at different stages of human development*

β-like chain	α-like chain	Haemoglobin present	Stage at which synthesised
ε	ζ	$\zeta_2\,\varepsilon_2$	early embryo
$^{A}\gamma\ ^{G}\gamma$	α	$\alpha_2\,\varepsilon_2$ } $\zeta_2\,\gamma_2$ } $\alpha_2\,\gamma_2$	foetus
β δ	α	$\alpha_2\,\beta_2$ } $\alpha_2\,\delta_2$ }	end of foetal development onwards

of evolution because there must have been duplications of genes coding for primordial forms of these globins. Subsequent mutations have given rise to the present-day forms which have some survival value, so the mutations have become fixed.

The DNA containing the globin genes occurs in two clusters (Fig. 9.2). The β-like genes, on chromosome 11, are spaced out over a length of about 60 kbp, and are highly homologous with each other. They all possess a common pattern of three exons with two introns in exactly corresponding sites (Table 9.2). The α-like gene cluster, on chromosome

Fig. 9.1. Critical bases and spacers in the 5′-flanking region of some of the globin genes. The numbers over the lines designate the number of bases in these positions. del = deletion. AC is the cap site, where transcription starts: ATG is the initiation codon, which has mutated in ζ_2.

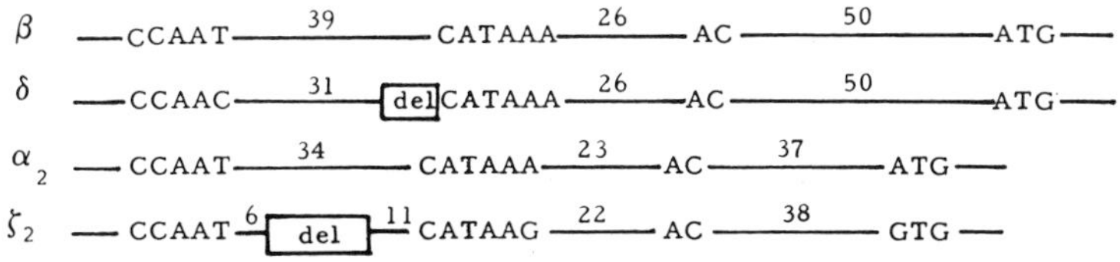

β —CCAAT— 39 —CATAAA— 26 —AC— 50 —ATG—
δ —CCAAC— 31 —[del]CATAAA— 26 —AC— 50 —ATG—
α_2 —CCAAT— 34 —CATAAA— 23 —AC— 37 —ATG—
ζ_2 —CCAAT— 6 —[del]— 11 —CATAAG— 22 —AC— 38 —GTG—

Fig. 9.2. Maps of the β- and α-globin gene clusters, on chromosomes 11 and 16 respectively, in the human genome.

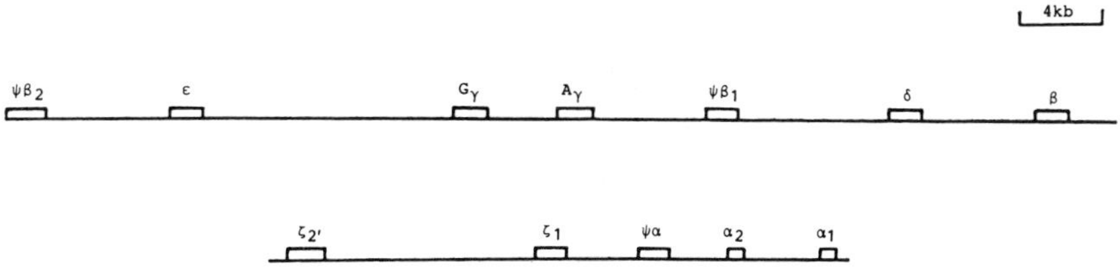

4kb

ψβ2 ε Gγ Aγ ψβ1 δ β

ζ2′ ζ1 ψα α2 α1

Table 9.2. *Organisation of the α- and β-globin genes in the human genome*

	Cap site to initiation codon	First exon	First intron	Second exon	Second intron	Third exon
α	38	96	127	204	133	126
β	54	92	138	223	889	123

16, is shorter (rather less than 30 kbp), and again the individual genes have a common pattern of three exons and two introns, though the introns are shorter than those in the β-like genes. In each cluster the genes are arranged from 5′ to 3′ in the same order as their temporal expression.

Gene duplications occurring recently in evolutionary history are likely to have given rise to the $^{A}\gamma$ and $^{G}\gamma$ pair, and also the δ and β pair. There is no evidence for a duplication of the ε gene, though there is ample space for a duplicated gene 3′ to the existing functional one. There has also been duplication of the α-like ζ gene, and triplication of the α gene itself. In the latter case, two of the genes present are identical, while the third α gene and one of the ζ genes have undergone mutations resulting in the production of non-functional pseudo-genes.

Pseudo-genes have DNA sequences which are highly homologous to those of transcribed genes, but which do not direct the synthesis of any functional polypeptides. Many pseudo-genes have been found, including two in the β-globin gene cluster. Pseudo-genes may contain insertions or deletions which cause premature chain termination of transcription products, or there may be mutations at the intron-exon junctions so that correct splicing cannot occur.

The pseudo α gene has mutations at the splice sites of both introns which make it unlikely that correct splicing could take place. Even if it did, a deletion of 23 bases in the second exon throws the reading frame out of phase so that a termination codon appears in phase. The pseudo ζ gene has a mutation giving rise to an in-phase termination codon in the first exon. It also has considerably longer introns than the true ζ gene. Introns in both ζ genes have numerous repeated sequences.

The globin genes of other species, such as the mouse and chicken, have also been studied and, in general, present a similar picture of developmentally regulated genes.

9.2 Thalassaemias

These are a collection of diseases in which the synthesis of one type of the globin chains is either reduced or absent. They are called α or β thalassaemias depending on the chain whose synthesis is deficient. These defects result in the production of a relative excess of the chain whose synthesis is not affected, and unusual tetramers (i.e. β_4 or α_4) are found which tend to cause destruction of erythrocytes or red cell progenitors. Consequently anaemia develops and when the condition is homozygous, affected individuals are likely to die early unless treated with repeated blood transfusions. These diseases are widespread in the warm regions of the world. The heterozygotes have an advantage over the normal population because they are more resistant to the malarial parasite.

Examination of the globin genes of a number of patients has shown that many different mutational events can give rise to the thalassaemias. In some cases, point mutations in the β chain gene have converted amino acid codons to termination codons (e.g. AAG (Lys) $\rightarrow$ UAG at position 14, or CAG (Gln) $\rightarrow$ UAG at position 39). mRNAs produced from these mutated genes will obviously direct the synthesis of shortened proteins, which do not pair correctly with the α chain and are rapidly degraded.

In another type, common in SE Asia, a mutation has occurred in the termination codon of the α chain (UAA $\rightarrow$ CAA). Since another in-phase termination codon is reached 90 bp downstream the mRNA that is produced codes for a longer than usual α chain. In fact, only very small amounts of this elongated chain are produced owing to the marked instability of the abnormal mRNA.

Other types result from mutations in or near the splice sites at which the introns are normally removed. A point mutation in codon 24 in the β chain leads to a pre-mRNA which is spliced incorrectly. This causes premature termination since the splicing generates a reading frame different from that used in the normal mRNA. In addition, the pre-mRNA is processed more slowly. A mutation in the first intron has been discovered which generates a new acceptor splice site 5′ to the usual one, and this also leads to premature termination because the reading frame is now out of phase. In other thalassaemias there are deletions of either the whole or 3′-portions of the β gene which lead to loss of ability to synthesise normal globin chains. Some of these mutations leading to β-thalassaemia are shown in Figure 9.3.

Some very interesting cases have been studied in which the affected individuals produce chains containing the N-terminal portion of the δ chain coupled to the C-terminal portion of the β chain. Analysis of the DNA shows that crossing over has occurred with fusion of the corresponding parts of the two genes (Fig. 9.4). This is believed to have occurred at meiosis. These hybrid δ-β chains are synthesised at a slower rate than the α chains so that an excess of the latter appears and a mild thalassaemia results. This must be because of the relatively inefficient initiation of transcription of the δ-β chain gene, particularly in homozygotes. These hybrid chains are known as Lepore chains, named after the patient in whom they were originally discovered. There are several chemically distinct Lepore haemoglobins in which the crossover has occurred in different positions. Anti-Lepore chains are also known with the reverse crossover, i.e. the N-terminal part of the β chain coupled to the C-terminal part of the δ chain. These cause hardly any symptoms of thalassaemia because there should be more efficient initiation of these particular hybrid genes.

9.3 Other mutations

There are many forms of mutant haemoglobin in which single amino acid substitutions have occurred in either α or β chains. The vast majority of these are due to single base changes in one of the codons in the DNA. Some of these abnormal haemoglobins have undesirable properties, and may lead to various forms of anaemia. The best known occurs in the condition of sickle cell anaemia in which there is a change of glutamate to valine in position 6 of the β chain, caused by a point mutation of the codon GAG to GTG. This small change in amino acid sequence leads to a profound difference in the properties of the haemoglobin that is present in affected individuals. The deoxygenated form precipitates out very readily and may damage the red cell membrane leading to haemolysis and

Fig. 9.3. Mutations (numbered) leading to various forms of β-thalassaemias. The sequence of the normal β gene is at the top with the amino acids which it encodes and their positions in the peptide chain shown above. Mutated nucleotides are shown below together with the amino acids specified by the mutated genes. d = deleted nucleotide.

```
 14          24  25  26  27   28  29   30                                     31  32
 Lys         Gly Gly Glu Ala  Leu Gly  Ar. . . . . . . . . . . . . . . . . . g Leu Leu
 AAG - - - - GGT GGT GAG GCC CTG GGC AG GTGG - - - - - - - GGTCTATTTTCCCACCCTTAG G CTG CTG - - -

 UAG         GGA G GTGAG - - - - - - - - - - - - - - - - - - - - - - - - - AG GC TGC TGG
 Ter         Gly                                                            Gly Cys Trp

  1           2                                     AG T CTA TTT TCC CAC CCT TAG
                                                    3     Leu Phe Ser His  Pro Ter

       39  40  41  42  43              57  58  59  60  61
       Gln Arg Phe Phe Glu             Asn Pro Lys Val Lys
 - - - CAG AGG TTC TTT GAG - - - - - - AAC CCT AAC GTG AAG - - - - - - - - GT - - - - - -

2 - - - - - - - - - - - - - - - - - - - CAA CCC TAA                        AT
                                        Gln Pro Ter                        no
4 - - - UAG                                                                splice
        Ter                                                                  6

5 - - - CAG AGG ddd d TTG AGT - - - - - - - - - - - -ACG TGA
        Gln Arg        Leu Ser                       Arg Ter
```

subsequent anaemia. This condition is quite widespread, especially among negroes, since the heterozygotes again have increased resistance to the malarial parasite. Anaemia is much more serious in the homozygotes.

Several hundred other single amino acid substitutions are known, and very many of them do not give any observable pathological effects. However, if a substitution occurs in a position of the globin which is critical for its functions (e.g. involved in binding haem, diphosphoglycerate, or with another subunit) deleterious effects may be observed. In

Fig. 9.4. Formation of Lepore and anti-Lepore globins by crossing over between the β and δ genes. The normal genes are shown in the middle, and may pair incorrectly because of their extensive homology.

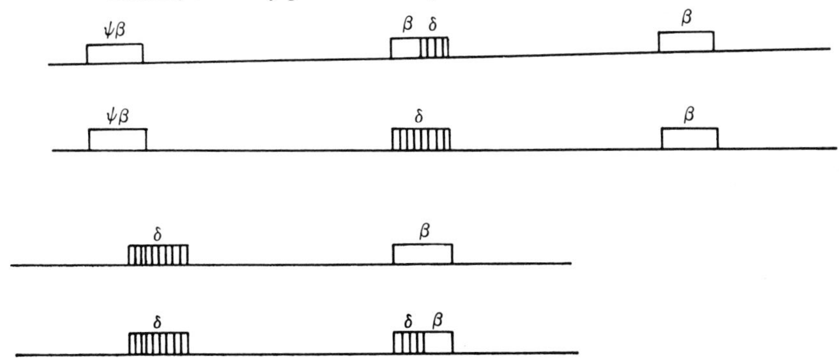

Fig. 9.5. Deletion of bases in β-globin Niteroi. This is believed to have occurred by mis-pairing of complementary bases following endonucleolytic chain scission and reformation.

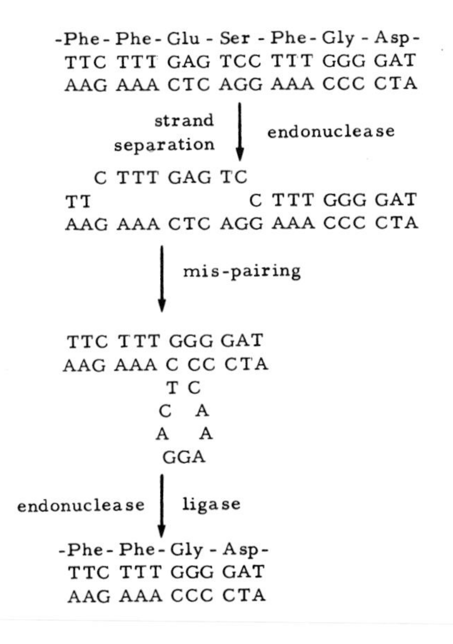

a few cases one or more amino acid residues are deleted. It is likely that these have occurred by unequal crossing over since they are found at sites where there are repeats of several bases in the sequence of the DNA. At these sites mispairing of adjacent sequences after a chromosomal break, followed by degradation of an unmatched sequence, could occur (Fig. 9.5).

Finally, there are a number of haemoglobins with C-terminal extensions of either α or β chains. These have arisen either as a result of a mutation in the termination codon (see earlier), or because of a deletion or insertion near this site. Deletions or insertions occurring in the α gene (Wayne) or in the β gene (Cranston) have lead to frame-shift mutations which bring into use new termination codons several residues 3′ to the usual ones (Fig. 9.6). These particular variants do not seem to have any undesirable consequences.

9.4 Prenatal diagnosis of anaemias

Prenatal diagnosis of genetically determined forms of anaemia is now technically feasible. Foetal cells can be obtained by amniocentesis and grown on to provide enough DNA for mapping by restriction endonucleases. This can often show up the presence of mutations leading to the expression of thalassaemic or other deleterious genetic changes (see

Fig. 9.6. Deletion and insertion of bases near the termination codon leading to lengthened α chains (Wayne) and β chains (Cranston). The underlined bases have been inserted in Cranston chain.

αCHAIN	Ser UCC	Lys AAA	Tyr UAC	Arg CGU	Ter UAA		
WAYNE	UCC Ser	AAU Asn	ACC Thr	GUU Val	AAG Lys	CUG Leu	GAG Glu
	CCU Pro	CGG Arg	UAG Ter				
βCHAIN	Lys AAG	Tyr UAU	His CAC	Ter UAA			
CRANSTON	AAG Lys	<u>AGU</u> Ser	AUC Ile	ACU Thr	AAG Lys	CUG Leu	GCU Ala
	UUC Phe	UUG Leu	CUU Leu	UCC Ser	AAU Asn	UUC Phe	UAU Tyr
	UAA Ter						

Chapter 3.11). Parents can then be offered the chance of termination of a pregnancy which would result in the birth of a severely handicapped child. Screening of large populations in areas where thalassaemia or sickle cell anaemia is common may not at present be economically practicable, but this situation could change as simpler techniques are developed for this purpose.

10

Proteins of the immune system

10.1 Immunoglobulins consist of H and L chains

Immunoglobulins (Igs), which are *antibodies*, are proteins consisting of four peptide chains – two identical light (L) chains of about 220 amino acid residues, and two identical heavy (H) chains containing between 450 and 600 amino acid residues. Each L chain consists of two *domains* of approximately equal size. The N-terminal domain is variable in amino acid sequence, and is different in each individual L chain that has so far been sequenced. The C-terminal domain is constant in sequence, though there are two types that exhibit a high degree of homology, known as κ and λ. Either one (but not both) may be found in one Ig molecule.

The H chain contains four or five domains, each with about 110 residues. Again, the N-terminal domain is variable in sequence, and participates with the N-terminal variable domain of the L chain in forming the antigen-binding site, of which there are two per molecule of Ig (Fig. 10.1). A short region, generally about 20 amino acid residues long, between the second and third domains, is called the hinge region. It is devoid of secondary structure, and therefore very flexible. There are five types of H chain – γ, α, μ, δ, ε – whose constant portions are quite distinct though homologous (but μ and ε have an extra domain and no hinge region). There are subclasses of γ and α chains. Intact Ig molecules are identified by suffixes of Latin letters corresponding to the Greek letter of the H chain (G, A, M, D, E). Variable and constant domains of the L chains are designated Vκ or Vλ and Cκ or Cλ. Corresponding designations of the domains of the H chains, starting from the N-terminal end of the molecule are V_H, C_{H1}, C_{H2} etc. The chromosomal location of the L and H chain genes in mouse and humans is shown in Table 10.1.

The constant domains of all H chains and both classes of L chain are recognisably homologous, and it is also possible to discern weak homology

Table 10.1. *The chromosomal location of immunoglobulin genes*

Protein chain	Gene symbol	Mouse chromosome number	Human chromosome number
H	Igh	12	14
λ	Igl	16	22
κ	Igk	6	2

of these domains with the V_H and V_κ and V_λ domains. This suggests that the present-day molecules have evolved by repeated duplications and mutations from a primordial Ig-like domain.

When an animal starts producing antibodies after immunisation, the first ones to appear in the plasma are IgM, but as the immune response runs its course there is generally a switch to other classes (IgG, IgA, etc) which retain the original antigen-binding specificity, and therefore the same V domains.

Antibodies are highly specific molecules in that each one will only recognise and bind to one (or a very limited number of) chemically defined groupings in an antigen molecule. The total number of antigens which can be recognised is extremely large – estimates of the order of 10^5 to more than 10^6 different specificities seem quite reasonable. A central problem of immunology is to explain how so many antibody molecules, all of the same general pattern, can be synthesised with different specificities, which are presumably manifested through variations in the amino acid sequence.

In both H and L chains the sequence variability is mainly confined to

Fig. 10.1. Structure of human immunoglobulin IgGl. The L chains are on the outside, joined to the H chains by disulphide bonds. The H chains are also joined together by a disulphide bond. The dark bands are the hypervariable regions. V_L and V_H are the variable domains: C_L, C_{H1}, C_{H2}, C_{H3} are the constant domains: h is the hinge region. There are two antigen-binding sites, each made up of the hypervariable region of one H and one L chain.

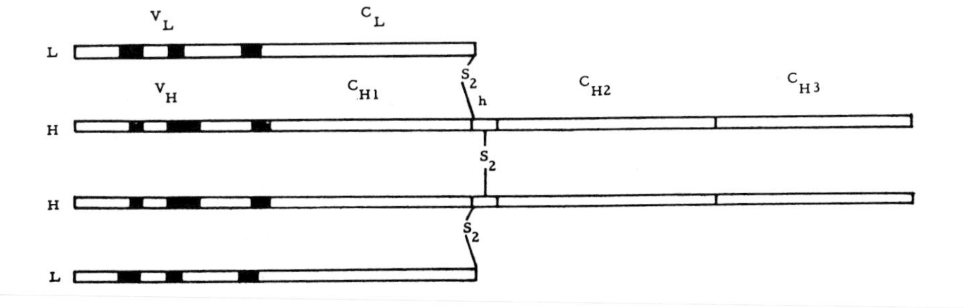

three regions, each containing from 5 to 16 amino acid residues, that are known as *hypervariable* or complementarity-determining regions. They are situated close together in the three-dimensional structure of Igs, forming a pocket where the antigen binds. Intervening stetches of the peptide chain are referred to as *framework regions*, and hold the hypervariable regions in the appropriate conformation for antigen binding.

Igs are synthesised in plasma cells which are the end products of the differentiation of B-lymphocytes. In mature animals there are very many clones of lymphocytes and the cells descended from them. The cells of each clone synthesise and secrete a single species of Ig possessing just one of the many possible V domains of both heavy and light chains, so the Ig fraction of plasma contains a highly heterogeneous population of many individual Ig molecules produced by all these clones. Injection of mineral oils into mice may cause a single clone of plasma cells to proliferate and form a plasmacytoma that produces a single homogeneous specific type of Ig. Plasmacytomas can be cultured through many generations of mice by repeated inoculations of cells from affected animals. This condition is analogous to the disease of myelomatosis in humans. Studies of the proteins produced in these conditions have yielded much information about the detailed structure of Igs.

10.2 The variability of immunoglobulins is due to genomic rearrangements

The variability in the primary structures of the polypeptide chains of the immunoglobulins arises mainly from rearrangements of lengths of genomic DNA that code for part of the V region of the H and L chains. In both cases there are a large number of exons coding for the first 90 or so amino acid residues from the N terminus. There are also four or five much smaller exons encoding 13 amino acids at the C-terminal end of the V domains, known as J (joining) genes. In the case of the H chains only, there are also a number of D (diversity) genes, situated between the V and J genes, that encode from 6 to 17 amino acids. During the maturation of B lymphocytes one each of the V, D and J genes of the H chains are joined together in a random fashion. In addition, insertions and deletions of nucleotides may occur during these recombinations. Nucleotides added in this way by the enzyme deoxynucleotidyl terminal transferase are known as N nucleotides. A very large number of different polypeptides are generated by these means. Prior to the synthesis of the mature L chains later during the ontogeny of the B cells there is random joining of their V and J genes. The best estimates for the numbers of all these events are

Table 10.2. *Possible number of human immunoglobulins*

Chain	H	κ	λ
V exons	200	100	6
D exons	> 20	—	—
J exons	6	4	6
Possible number of polypeptides	24 000	400	6
Possible number of proteins		$> 10^6$	

No account has been taken of extra proteins produced by the use of N nucleotides.

shown in Table 10.2. Since any H chain can be joined with any L chain, a very large number of different immunoglobulins are generated in this way. As the immune response runs its course, somatic mutation in the region of the genes coding for the hypervariable regions gives rise to an even larger diversification of specificity. This tends to produce antibodies with a higher affinity for the immunising antigen.

10.3 L chain genes

There are probably between 100 and 300 Vκ genes in the mouse: there may be only a rather small number in humans. They occur as families whose members have presumably been derived from a founding member by mutation. They are situated 5′ to the cluster of J genes (Fig. 10.2). Each V gene is preceded by a separate exon encoding the signal peptide.

The mouse produces very few λ chains and the arrangement of their genes is rather different from that of the κ chain genes. Of the few Vλ region genes, one is linked to either $J\lambda_3$ and $C\lambda_3$ or a $J\lambda_1$ and $C\lambda_1$ pair, while another is similarly linked to a $J\lambda_2$ and $C\lambda_2$ pair (Fig. 10.2).

The human locus encoding the λ chain is more complex and produces more types of this chain, but it has not been completely mapped.

10.4 H chain genes

Fig. 10.3 shows the structure of part of the H chain locus, with the exons for the μ chain. There are two forms of this chain with different sequences at the C-terminus that are produced by alternate splicing. μ_m is membrane bound with a predominantly hydrophobic C-terminal sequence that anchors it to the cell membrane where it acts as an antigen receptor after combination with an L chain. When a cell encounters an antigen that can bind to it, the complex is internalised and somehow directs the processing of the primary transcript to produce μ_s. This contains a short C-terminal extension to $C\mu_4$ consisting predominantly of hydrophilic amino acid residues and is secreted from the cell in combination with an L chain. In the genome the DNA encoding the C-terminal amino acid sequence of μ_s is directly adjacent to the 3′-end of the DNA coding for $C\mu_4$, while there are two further exons downstream encoding the C terminus of μ_m. Both the μ_s and the μ_m exons are followed by untranslated bases including the poly-adenylation signal. The choice between producing μ_s and μ_m depends on which poly-A site is used, since it is believed that poly-A addition precedes processing and splicing.

The δ gene is located about 2·5 kbp on the 3′-side of the μ gene, and both these genes are transcribed during the early part of the immune response, when IgD is found on the cell surface in addition to IgM, acting as an antigen receptor.

The genes for the constant regions of the other sub-classes show some similarity to each other (Fig. 10.4). There are generally three exons for the

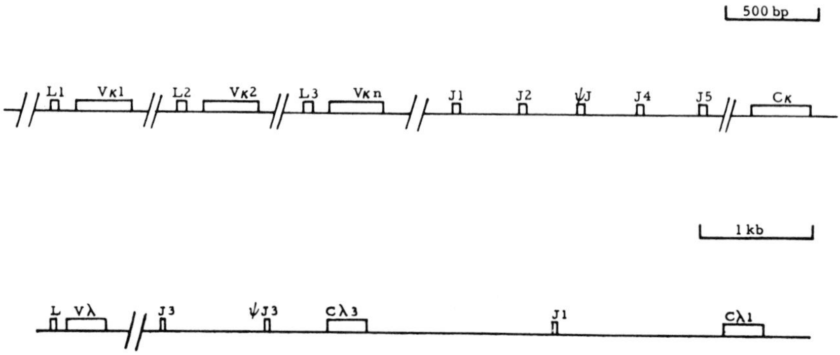

Fig. 10.2. Germ-line organisation of κ chain gene (top), and λ chain gene (bottom) in the mouse. There is a second similar cluster of λ chain genes. The distances between exons is not known where the lines are broken.

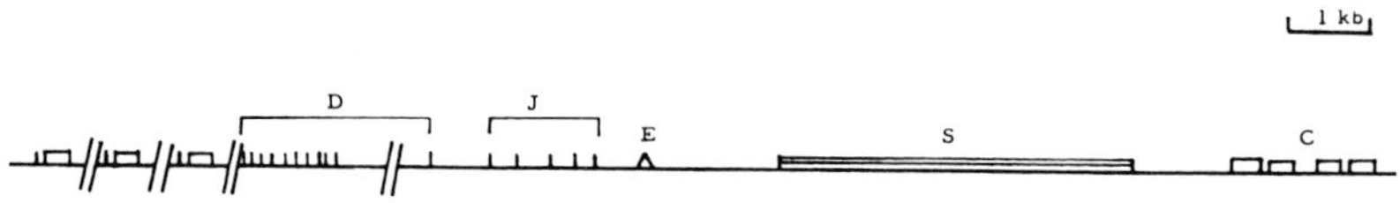

Fig. 10.3. Germ-line organisation of the H chain exons making up the V region in the mouse, showing also the Cμ exons.

major domains, and an exon for the hinge region, though in some cases this has been duplicated or quadruplicated, and sometimes it is joined to an adjacent exon for one of the domains. There are also downstream exons encoding membrane-bound forms of these proteins. There has been appreciable diversification during the evolution of the various classes of Igs. IgM is the most primitive one, since it occurs in fishes.

The exons encoding the C regions of the γ, ε and α chains are arranged somewhat differently in the mouse and human genomes (Fig. 10.5). This suggests that there has been a large duplication of that part of the genome

Fig. 10.4. Human C region genes. The open boxes are coding sequences, and the black boxes are untranslated sequences 3′ to the termination codons. Exons encoding the hinge regions are designated h. All the genes have additional downstream exons coding for amino acids that can insert into the cell membrane, but they are only shown for the μ chain. The single lines are introns.

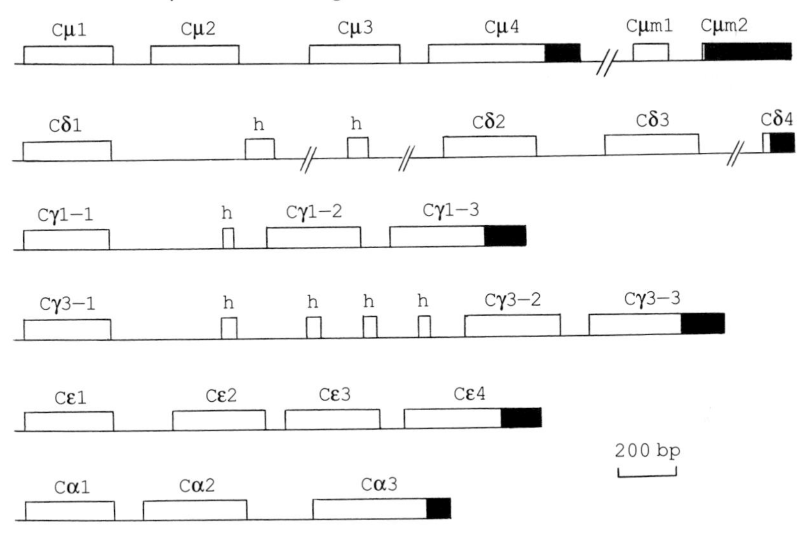

Fig. 10.5. Map of the C_H gene region in the mouse (top) and human (bottom). Individual exons are not shown, and the boxes showing the sizes of the exons are not drawn to scale.

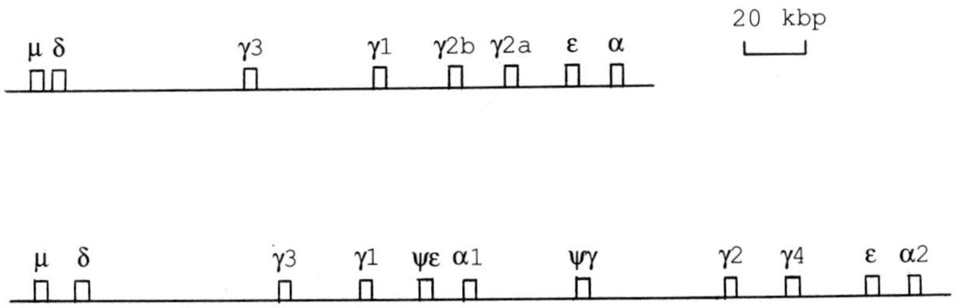

20 kbp

μ δ γ3 γ1 γ2b γ2a ε α

μ δ γ3 γ1 ψε α1 ψγ γ2 γ4 ε α2

originally containing one (or possibly two) γ genes, an ε gene and an α gene during the descent of the human race, while in the lineage leading to the modern mouse it is only the γ genes that have been duplicated.

10.5 DNA processing is employed during the course of the immune response

Three types of nucleic acid processing are involved in the expression of Ig genes. First, there is joining of V and J genes (and D genes in the case of the H chains). These genes are widely separated, and become juxtaposed in Ig-producing cells as the result of rearrangements of the DNA, brought about by the pairing of complementary sequences located at the 3′-ends of V exons, at the 5′-end of J exons and flanking both ends of D exons.

There are highly conserved heptamers (CACTGTG) and nonamers (GGTTTTTGT) 5′ to the end of the D and J exons that are complementary to sequences 3′ to V and D exons (Fig. 10.6). The gaps of 12 or 23 bp between the heptamer and nonamer correspond very nearly to one or two turns of the DNA helix so that these two sites are on the same face of the molecule and reasonably close to each other. This alignment provides sites for an enzyme to bind and remove all the sequences between both the V and D and the D and J exons, with subsequent ligation of the cut ends. The heptamer and nonamer separated by 23 bp always pair with the corresponding oligomers separated by 12 bp, ensuring that V exons are always joined to D exons and D exons to J exons.

Transcription of these processed DNA sequences yields mRNAs with a small intron near the end of the sequence coding for the leader peptide, and a much larger one between the J exon and the first C exon. These introns are excised in the usual way. Thus, the final production of the mature mRNA requires processing at the level of RNA rather than DNA.

Fig. 10.6. VDJ joining, 9 and 7 represent the highly conserved nonamers and heptamers which can pair and align the DNA for the process of splicing.

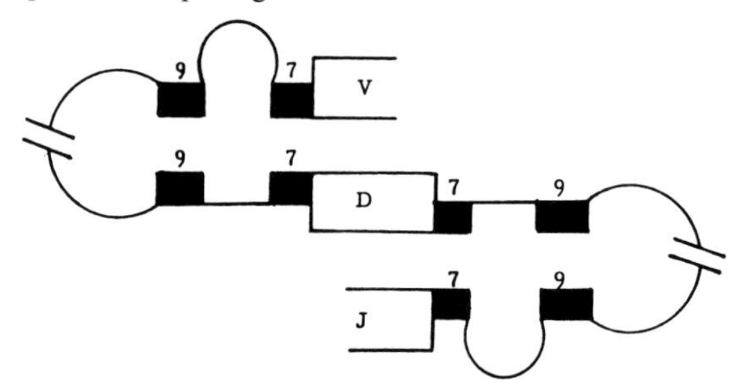

Rearrangements of the H chain genes at the DNA level are also involved in switching from the synthesis of one class of H chain to another. Several switch sites are located in the untranslated stretch between the 5′-end of the Cμ gene and the J cluster in the region shown as S on Figure 10.3. Different sites are used for joining to the C_λ, C_ε, and C_α genes, and there are no obvious homologies in these base sequences. The sites so far identified are clustered just 5′ to the end of a stretch of 2.6 kbp consisting almost entirely of two highly conserved pentamers – GGGGT and GAGCT.

There are highly conserved tandem repeats of 49–80 bases 5′ to the C_λ, Cε and C_α genes containing the two pentameric sequences that also occur in the switch sites. These repeated base sequences are most probably involved in the recognition by the processing enzymes that are activated when switching from the synthesis of one class of Ig to another occurs.

There are no switch sites on the 5′-side of the Cδ gene which is situated only 2.5 kbp to the 3′-end of the Cμ gene. Simultaneous expression of μ and δ chains occurs before B lymphocytes differentiate into plasma cells. Alternative modes of processing a pre-mRNA lead to the production of both μ and δ chains.

10.6 Transcriptional control of Ig genes

In the 5′-flanking region of Ig L chain genes about 100 nt upstream from the site of initiation of transcription is a highly conserved octamer ATTTGCAT. A sequence inverted and complementary to this occurs in a similar position in H chain genes. A protein (NF-A2 or OCTII) that binds specifically to this sequence and stimulates transcription is expressed only in lymphoid tissues, while a different protein (NF-A1) that also binds to this sequence is more widely distributed.

There are enhancer sequences within the introns between the J and C exons of both κ and μ genes. These reside in 140 bp sequences that can be engineered into vectors derived from the SV40 genome which is then able to express genes that would not otherwise be expressed. These enhancers are tissue specific, only functioning in lymphoid cells, but not in fibroblasts or liver cells. Their sequences are similar to the core sequences of viral enhancers (Chapter 7.14).

These enhancers are probably required for the expression of rearranged H and L chain genes. The promoter sequences 5′ to the V genes may only be able to function when they have been brought near enough to the enhancers situated downstream from them in the rearranged gene. Thus any un-rearranged genes are not able to be transcribed.

A nuclear protein, NF-κB, that binds to these sequences, is only expressed in lymphoid tissues, though similar proteins occur elsewhere.

NF-κB also binds to enhancer sequences in other genes such as those encoding interleukin-2 and the mouse MHC antigen H-2K (see Chapter 10.9). A surprising feature of the interaction of this factor with the cognate enhancer sequence is that it is strongly stimulated by nucleoside triphosphates. It confers inducibility by TPA (see p. 113) on these genes and is found in two forms. An inactive form occurs in the cytosol of cells complexed with an inhibitory protein I-κB from which it can be liberated by treatment with deoxycholate (DOC) – an agent that can dissociate proteins that are physically, but not chemically, bound together. The other form is found in the nuclei after TPA stimulation and does not require activation by DOC. TPA treatment activates protein kinase C which in turn probably causes phosphorylation of I-κB leading to its dissociation from the complex with NF-κB. This mechanism is reminiscent of the activation of the glucocorticoid receptor which is normally found in the cytosol associated with the 90 kD heat shock protein in a tight complex. The binding of a glucocorticoid causes the complex to dissociate, leaving the receptor free to migrate to the nucleus (Chapter 7.10). It will be interesting to learn whether this is a general mechanism for activation of transcription by other stimulatory agents such as cAMP.

10.7 Allelic exclusion

Since all somatic cells possess pairs of homologous chromosomes, it might be expected that a fully differentiated lymphocyte would produce two types of Ig – one programmed by the genes on each chromosome. However, this never seems to happen, and only one of the pair of chromosomes directs the synthesis of Ig. This phenomenon is known as allelic exclusion, and has been investigated extensively. In some cases (more common with L chain genes) only one copy of the chromosomal DNA has been rearranged to contain the complete L chain gene ready for transcription. In other cases (more frequent with H chain genes) both copies are rearranged, but one undergoes an abortive rearrangement leading to inability to produce a functional polypeptide. In some cases proteins are secreted that are presumably products of mis-arranged genes, in the sense that they lack those amino acid sequences that are normally coded for by a separate exon.

10.8 T-cell receptor

There are two major classes of lymphocytes in the body, identified on the basis of different functions and by bearing different sets of surface antigens. B-lymphocytes mature in the bone marrow (or bursa of Fabricius in birds) and are the precursors of plasma cells that synthesise

Table 10.3. *Possible specificities of T-cell antigen receptors*

	α	β	γ	δ
V exons	100	25	7	10
D exons	0	2	0	2
J exons	50	12	2	2
Possible intra-chain combinations	5000	600	14	40
Possible inter-chain combinations	3×10^6		560	

The numbers of intra-chain combinations is greatly increased by the use of N nucleotides.

Igs. T-lymphocytes mature in the thymus and are involved in antigen-specific reactions leading to damage of other cells containing the particular antigen, or to secretion of lymphokines required for the differentiation of B-lymphocytes through various stages to plasma cells. Among their cell surface antigens are the T-cell receptors – antigen-specific molecules, parts of which have some homology with Igs, but which are never found free in solution.

There are two kinds of T-cell receptor, each made up of two rather similar polypeptide chains, encoded by separate genes that are somewhat analogous to those of Igs. There are leader and V exons, D exons (in some cases) and J exons that must be rearranged to produce the functional protein. The remainder of the genes contains an exon encoding a domain homologous to one of the constant region domains of immunoglobulins, a small exon encoding a connecting peptide and a larger one encoding the transmembrane segment and a short cytoplasmic tail.

The T-cell receptors on nearly all circulating T-lymphocytes are made up of α and β chains, while in certain parts of the body (e.g. dermal and intestinal epithelia) they contain γ and δ chains. Other combinations have not so far been found and are not believed to occur. Table 10.3 shows the approximate numbers of the various segments of the T-cell receptor genes that can be assembled in a combinatorial fashion to yield the mature genes. Note that the V–D–J combinations are again made at the level of DNA as in the case of the Igs. In some cases, during these recombinations, additional N nucleotides, not encoded in the genome may be incorporated between the V and D and between the D and J segments by the action of the enzyme deoxyribonucleotidyl terminal transferase. This further increases the diversity of the T-cell receptor proteins, but if the number of nucleotides added is not a multiple of three, the sequence to the 3′-side of the addition will be out of the normal reading frame and so no functional protein can be made. Cells in which this occurs probably die during development in the thymus.

The location of the exons coding for the δ chain is unusual, since their constituent parts are found between the exons coding for the J and C regions of the α chain. There is also a C region exon of the δ chain oriented in the opposite direction to the rest of the exons. It must undergo an inversion when the gene is re-arranged.

10.9 The major histocompatability complex

When a tissue (other than erythrocytes) is transplanted from one individual to another the graft will generally survive for only a few days. This is because there are certain integral membrane proteins on the surface of all nucleated diploid cells, encoded by highly polymorphic genes that allow 'self' to be distinguished from 'non-self'. These transplantation antigens control, in various ways, co-operation between different types of lymphocytes in the immune system when it has been stimulated by molecules foreign to the organism. Their discovery, through work on tissue transplantation, was a serendipitous accident. Because of their major clinical importance they have been investigated widely in mice (as

Fig. 10.7. Map of Major Histocompatibility Complex on chromosome 6 (top), and mouse H2 complex on chromosome 17 (bottom). The position of known genes are marked by vertical lines above the chromosome. The lines below the human chromosome are open reading frames for which no protein product has yet been detected. The lines below the mouse chromosome indicate the limits of a deletion in some strains leading to loss of 3 D genes and 1 L gene. The Class III genes encode members of the complement family and steroid hydroxylase C-21. TNF = Tumour Necrosis Factor: hsp = heat shock protein.

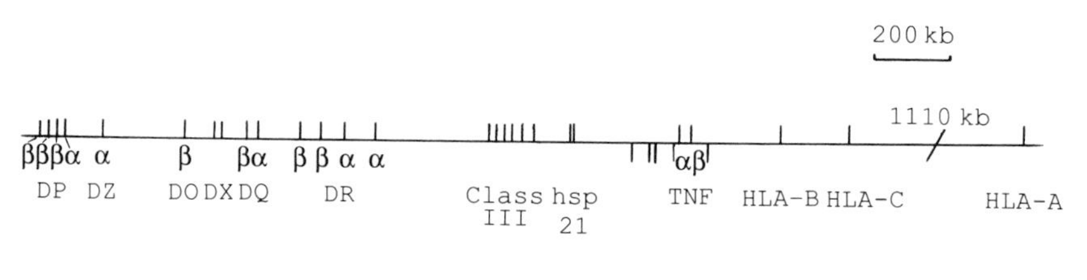

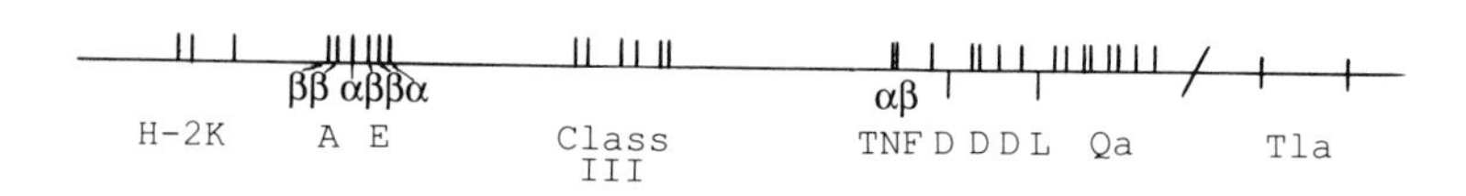

a convenient model system), and in humans. The antigens are part of the Major Histocompatibility Complex (MHC) that is encoded on chromosome 6 in humans. In mice this is known as the H-2 complex and is encoded on chromosome 17 (Fig. 10.7). These complexes extend over two to four million bp and contain genes encoding two distinct groups of cell-surface antigens as well as some other proteins.

Class I molecules are all similar in structure and are non-covalently but firmly bound to a small protein called β_2-microglobulin (Fig. 10.8). The classical transplantation antigens (K, L, D in the mouse; HLA-A, HLA-B, HLA-C in humans) are all highly homologous in sequence. The proteins consist of three extracellular domains of about 90 amino acid residues each, a transmembrane domain with a hydrophobic core, and a cytoplasmic domain. All three are expressed as integral membrane proteins on all nucleated diploid cells, but are found in much larger amounts on the surface of all kinds of lymphocytes. They have important roles in events leading to the killing by cytolytic T-cells of cells bearing non-self antigenic determinants.

There is a second group, designated Qa and Tla in the mouse, that are found mostly on lymphocytes, particularly those of the T-lineage that have matured in the thymus gland. Human homologues have been detected serologically but have not been characterised; nor have their

Fig. 10.8. Structure of the Class I (right) and Class II (left) transplantation antigens showing the domain structure. S_2 = disulphide bonds. β_2m = β_2-microglobulin. The cell membrane is in the middle between the dotted lines with some of the polypeptide chains passing through it into the cytoplasm at the very bottom.

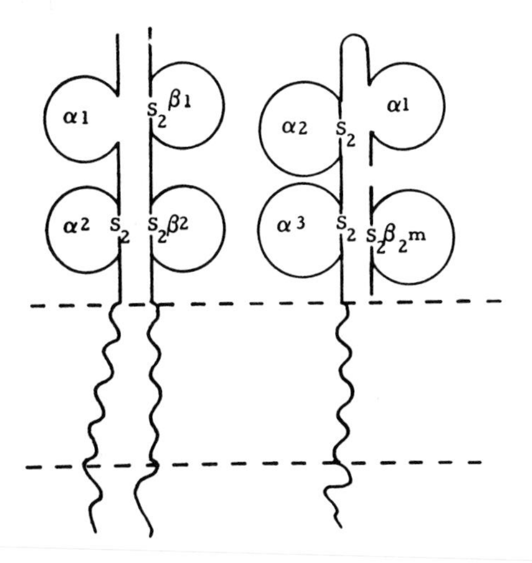

genes been mapped. These antigens may be involved in cellular differentiation since they are found on certain embryonic tissues. Their mRNAs are present in a wide range of tumour cells. Malignant cells often display many characteristics of undifferentiated cells, so it is understandable that they should express the genes encoding these antigens.

The MHC also encodes a second set of integral membrane proteins (Class II) whose structure is somewhat reminiscent of the Class I proteins. The Class II proteins contain two non-covalently associated polypeptide chains but are not complexed with β_2-microglobulin. Each chain has transmembrane and cytoplasmic domains but there are only two extracellular domains (Fig. 10.8). However, in each case one of the domains (α_2 and β_2) is homologous to a constant region domain of the immunoglobulins, and the other extracellular domain is of a similar size. In the mouse there are two sets – designated I–A and I–E – while in humans their counterparts are called DQ and DR, and there are two others called DP and DO/DX, for which no murine homologues have been found. Their distribution is limited to cells of the lymphocyte and monocyte lineages. The genes encoding them were formerly called the Immune Response (Ir) genes, since their products control the magnitude of this response by binding and presenting fragments of antigens to T and B lymphocytes.

10.10 Class I genes

These genes are highly polymorphic – about 50 alleles have been described at some loci in both mice and humans. Alleles in the mouse are designated by lower case superscripts (e.g. K^b or K^d; L^b; D^b, etc), while human alleles are given numerical designations (e.g. HLA-A2; HLA-B7; HLA-C5, etc). Each domain of the protein is encoded by one or more separate exons. In the mouse the cytoplasmic domain is encoded by three short exons, while in humans there are only two exons in this part of the gene (Fig. 10.9). Most of the sequence variation between alleles occurs in the first and second extracellular domains. The third extracellular domain and also the β_2-microglobulin have distinct homology with constant region domains of the immunoglobulins. β_2-microglobulin is invariant and is encoded on a different chromosome from the MHC antigens.

The polymorphism of these antigens resides in multiple substitutions which are clustered, rather than as individual substitutions resulting from point mutations. This suggests that many of the alleles in this system have arisen by gene-conversion events.

The H-2K and HLA-A genes are in very different positions relative to the other MHC genes in the two species. It is possible that the H-2K gene has been translocated since the divergence of the lines leading to humans and mice. Many strains of mice possess four distinct but similar H2-D antigen genes, but in certain strains the genes for three of the D antigens, the H2-L antigen and some of the Qa antigens have been lost.

The Qa and Tla genes are much less polymorphic and occur as clusters of distinct genes over a length of about 250 kbp (Fig. 10.7). So far, 10 genes have been mapped to the Qa region and 21 to the Tla region. It is not known how many of these are expressed (some may be pseudogenes). They could provide a reservoir of donor sequences for use in gene-conversion events to give rise to the multiple alleles found in the K, D and L genes.

10.11 Class II genes

The Class II polypeptides are encoded by genes with exons corresponding to each of the domains, though in some cases the trans-membrane and cytoplasmic axons are already joined, and in others more than one exon codes for the cytoplasmic domain.

The mouse I–A and I–E genes are located on a 200 kbp length of chromosome 17, encoding both α and β chains, and also two isolated single exons that are pseudogenes.

The corresponding portion of the human genome is about 1000 kbp long and contains all the genes for the Class II antigens and several pseudo-genes.

The α and β chains are transcribed in opposite directions, and in some cases there is more than one gene for either one or both of them. During the evolutionary time separating the murine and human lineages, this part of the genome must have been duplicated giving rise to the genes found in humans for which there are no homologues in the mouse.

The β chains are generally the polymorphic ones, though there is

Fig. 10.9. Structures of the H2-K^b gene of the mouse (top), and of a human HLA-B gene (bottom). L = leader peptide sequence: α_1, α_2, α_3 = extracellular domains: TM = transmembrane domain: CY = parts of the cytoplasmic domain: UT = untranslated sequences.

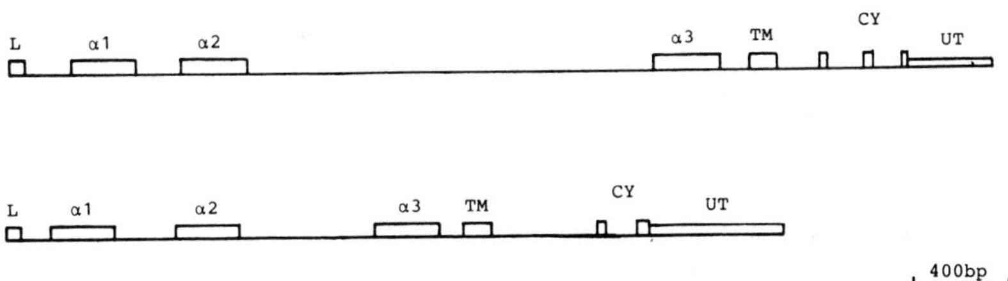

Table 10.4. *Selected members of the immunoglobulin superfamily*

	Number of Ig domains
Igκ, Igλ	2
IgH	4/5
T cell receptors	2
HLA-A, -B, -C	1*
HLA-DP, -DQ, -DR	1*
β_2-microglobulin	1
CD4	3
CD8	1*
poly-Ig receptor	5
n-CAM	5
Platelet derived growth factor	5

* These proteins have two polyeptide chains, each containing one Ig-like domain.

sometimes some polymorphism in the α chains, and the polymorphisms are confined to the β_1 and α_1 domains that are probably well clear of the cell surface. It is these domains that are recognised as self by other cells of the same species when they function in cooperative reactions between T- and B-lymphocytes and macrophages. There is no evidence for the transcription of several pseudogenes situated in the Class II locus encoding only incomplete α and β type polypeptides.

10.12 Ig superfamily

With the sequencing of many genes, including those encoding proteins involved in the immune response, it has become apparent that many of them are related to the better known Ig genes. So far, over 40 such genes have been identified. A large proportion of them code for integral membrane proteins found principally on lymphocytes and macrophages, but a few occur in other tissues such as brain and muscle. A selection of these genes comprising the Ig superfamily is shown in Table 10.4. Only the Ig and T-cell receptor genes undergo re-arrangement at the level of DNA, and they, and the MHC genes, are the only ones that exhibit a high degree of polymorphism. Several of the other genes encode more than one domain homologous to the domains of the Igs, but many have only a single Ig-like domain.

Since many of the proteins encoded by these genes are found on cell surfaces and are involved in various sorts of recognition and communication between cells, it has been suggested that these were the primary functions of members of this family. It was only later that molecules capable of recognising and combating foreign molecules evolved and led to the development of the immune system. So far, there is little evidence that genes of this family are found in organisms other than vertebrates, but this may be because there have been few investigations in this area.

10.13 Complement genes

The genes for some of the components of the complement system, that brings about the lysis of cells recognised as foreign to the organism, are found in the middle of the MHC (Fig. 10.7). Those encoding the second component of complement (C2) and a protein known as Factor B are extremely close together. These are both unusual serine proteases that have analogous functions in two pathways by which complement is activated. There are additionally two genes for the fourth component of complement (C4) that are not quite so closely linked.

In the mouse, only one of the C4 genes encodes a protein with haemolytic activity. The second codes for a very similar protein known as the sex linked protein (slp) that is only expressed in males. The two C4 genes in humans encode slightly different proteins with different haemolytic activities. They are highly polymorphic with 13 variants at one locus and 22 at the other but little is known of the precise differences between the alleles.

Adjacent to each C4 gene is a gene coding for steroid 21-hydroxylase – an important member of the cytochrome P450 family of enzymes – required for the synthesis of certain adrenal steroids. This enzyme has no connection with the proteins of the immune system and it is presumably found in this situation due to an accidental translocation. Only one of the genes is expressed, and if this is defective a congenital form of adrenal insufficiency and hyperplasia may result that can prove fatal if it is not treated.

Additional genes have been discovered within the bounds of the MHC in the human genome. These include two closely linked and similar genes encoding Tumour Necrosis Factor, also known as Cachectin (TNFα) and Lymphotoxin (TNFβ), and duplicated genes for one of the heat shock proteins. Several others with open reading frames have been described though none of the proteins that could be produced from them have yet been identified.

11

Some gene families

11.1 Collagen

Collagen is the most abundant protein in the body, comprising approximately 25% of the body's protein in humans. There are actually five distinct but closely related types. The molecules of each type consist of three long polypeptide chains tightly wound together to form a triple helix, so that the molecules are fibrous. In Types I, IV and V there are two identical chains and one that is different, while all three chains are identical in Types II and III. Altogether eight different polypeptides are synthesised and, as far as is known, each has its own gene. There are extensions at each end of the molecules (telo-peptides) that are cleaved off after the chains are wound together to form the triple helix. The mature chains are about 1000 amino acids long and are characterized by the presence of a glycine residue at every third position through nearly all their length. Its tiny side chain can be easily accommodated inside the tightly wound triple helix. They are also unusually rich in proline, and many prolyl residues (and also some lysyl ones) undergo post-translational hydroxylation. Exons tends to be rather short and the gene for the Type I contains no fewer than 54. They mostly contain 45 or 54 nucleotides or multiples or sums of these numbers, suggesting their evolution from a small coding unit. The introns also tend to be fairly short so that the genes are not inordinately long.

The genes for the two types of polypeptide chain of Type I collagen are of very different lengths although they encode proteins of very similar size. The α1I gene is 18 kbp long, of which approximately 12 kbp encode the introns, while the α2I gene is 38 kbp long with about 33 kbp encoding the introns. This means that if they are both transcribed and translated at the same rate twice as much of the α1I polypeptide as the α2I polypeptide should be produced. In fact, measurements of the amounts of the two mRNAs in fibroblasts show that this is so. This ensures that correct amounts of the precursors for the mature Type I collagen are synthesised, since it contains two α1I chains and a single α2I chain.

In general, the genes for the various collagen chains are dispersed throughout the genome, but those for the α1IV and α2IV are closely linked. In the mouse they are in a head-to-head orientation with their cap sites only about 270 bp apart so that their expression could be co-ordinately controlled in a fairly simple manner.

There are a number of heritable conditions in which abnormal and malfunctioning collagen molecules are made. These can result from loss of one or more exons or from point mutations in which a glycine codon is mutated to one coding for another amino acid. Both these changes interfere with the packing of the polypeptide chains into the tight and stable triple helix so that the collagen fibres are weakened.

A few other proteins contain collagen-like domains, characterised by the location of glycine at every third residue. The complement protein C1q contains three types of such chains whose overall structures are very similar. The only one whose gene has been sequenced (C1qB) does not have the typical collagen gene structure for the collagenous domain, since it only contains one intron which is situated at the junction between this domain and the rest of the molecule.

Collagens are also found in invertebrates, but the few genes so far sequenced in *Drosophila* and *Caenorhabditis* show little resemblance to those of vertebrates, since they contain much longer exons and fewer introns.

11.2 Cytochrome P450

Cytochrome P450 is the generic name for a large family of proteins that catalyse the oxidation of some normal metabolites and many foreign compounds (xenobiotics). These enzymes generally occur bound to the endoplasmic reticulum in eukaryotes, though some are found in mitochondria. They are encoded by a superfamily of genes that is made up of at least ten different families classified on the basis of differences in amino acid sequences. In the families for which genomic sequences are available there are marked differences in intron-exon structure (Fig. 11.1).

The members of family II all have unusually long exons of about 830 nt encoding the amino terminal end of the protein. In many of the enzymes this end of the molecule is rich in amino acids with hydrophobic side chains that make up the site of binding to the endoplasmic reticulum. Near the carboxyl terminus of the proteins there is a highly conserved region containing a cysteine residue that binds the haem moiety of the enzyme.

Fig. 11.1 Cytochrome P450 genes. The open rectangles are coding sequences; the filled ones are non-coding sequences; the connecting lines are introns. The figures above the slashes are the lengths (kbp) of introns that are too large to draw to scale. From top to bottom the proteins encoded are: human benzo(a)pyrene hydroxylase, human ethanol-inducible oxidase, human steroid side chain cleavage enzyme, human steroid 17-hydroxylase, human steroid 21-hydroxylase.

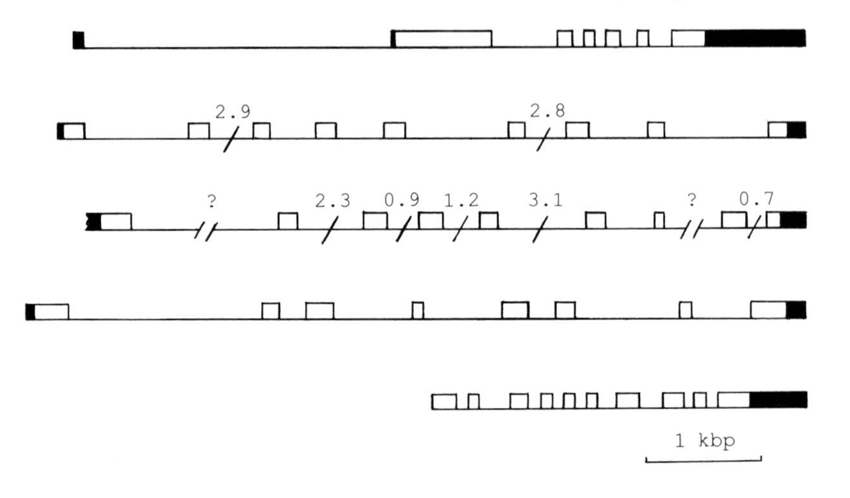

Fig. 11.2. Serine protease genes. Conventions as in Fig. 11.1. The letters H, D and S are placed above the positions where the histidyl, aspartyl and seryl residues of the catalytic triad are encoded. From top to bottom the proteins encoded are: trypsinogen, chymotrypsinogen, pro-elastase, cytotoxic cell protease, complement factor B.

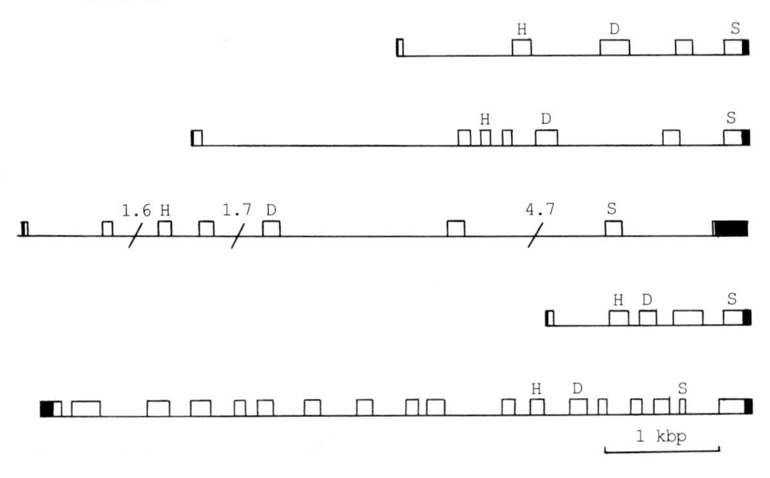

This is probably an ancient gene family since the bacterium *Pseudomonas putida* contains a homologous enzyme for oxidation of camphor. Some enzymes of this family have also been found in fungi, including yeast, but none have so far been reported in higher plants.

Some members of the family are produced constitutively, particularly those that catalyse the hydroxylation of steroids, since these are normal metabolic reactions. The transcription of their genes is often controlled by cAMP, acting as a second messenger for various trophic pituitary hormones. Other members of the family are inducible by various foreign compounds, particularly phenobarbitone which is itself metabolised by some of the enzymes whose synthesis it induces. Members of family I that oxidise polycyclic hydrocarbons are induced by these compounds after they have bound to a receptor protein known as Ah (for Aromatic hydrocarbon). It is believed that the complex of Ah and the hydrocarbon binds to sites 5′ to the gene and stimulates its transcription.

11.3 Serine proteases

The serine proteases constitute a large family of proteolytic enzymes whose catalytic activity always depends on a triad of histidyl, aspartyl and seryl residues. In most of these enzymes these three critical residues are encoded in different exons, but the distances between them are remarkably conserved. There are about 50 residues between the histidyl and aspartyl residues and between 90 and 108 (exceptionally 123) separating the aspartyl and seryl residues.

The simplest genes are those encoding the precursors for the well-known digestive enzymes trypsin, chymotrypsin and elastase (Fig. 11.2). The trypsin gene has five exons with the histidyl, aspartyl and seryl codons in exons 2, 3, and 5. There are several genes for other proteolytic enzymes with the same general structure, including a large family of about 25–30 very similar genes in the mouse encoding the kallikreins. These are closely linked on chromosome 7 and must have arisen by repeated duplications.

The chymotrypsin gene has an extra intron between those encoding the histidyl and aspartyl residues, while the elastase gene has an additional intron very near the 3′-end of its coding section.

The beginnings of the introns just downstream from the histidyl residue and the ends of the introns just upstream from the seryl residue are remarkably well conserved, though this is not the case with the intron/exon structures around the aspartyl residue. All these genes probably evolved from a trypsin-like gene into which additional introns have been inserted.

Some of the enzymes involved in the blood-clotting cascade are serine proteases, and exons encoding the essential catalytic amino acid residues are always found at the 3′-end of the genes (Fig. 11.3). Interestingly, in some of them the aspartyl and seryl residues are encoded in the same exon, suggesting that an intron may have been lost at some time. However, these are clearly examples of '*mosaic*' genes as they contain exons near the 5′-end encoding protein domains with definite discrete structures. These include domains found also in fibronectin that may be involved in binding the proteins to collagen and other tissue substrates; domains rich in glutamyl residues that are post-translationally modified by a vitamin K mediated reaction to yield carboxy-glutamyl residues that bind Ca^{2+}; small domains tightly internally bonded by disulphide bridges that are found in the epidermal growth factor; and domains so far unique to these proteins known as kringles that also contain disulphide bridges forming less compact structures.

There are also some serine proteases in the complement cascade. The genomic sequence of only one of these is known (factor B, Fig. 11.2). This

Fig. 11.3. Genes concerned with blood clotting. Conventions as in Fig. 11.1 and 11.2. The letters under portions of the genes indicate areas where there are γ-carboxyglutamyl residues in the mature proteins (G); sequences homologous to EGF (E) or fibronectin (F); kringles (K); repeats unique to factor XI (R). From top to bottom the proteins encoded are; factor VII, factor IX, factor XI, factor XII, factor V, tissue plasminogen activator, prothrombin.

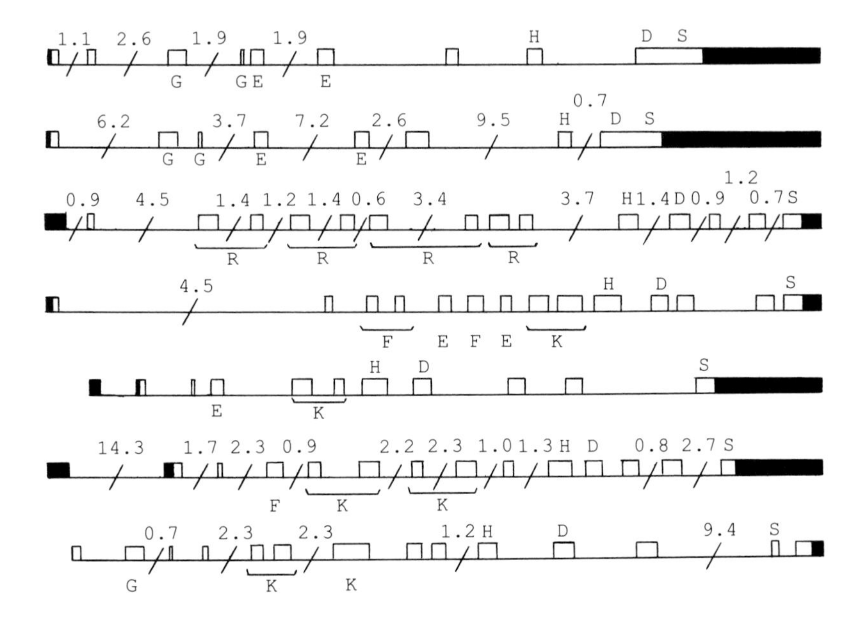

contains three exons encoding a domain of 60 amino acid residues (the SCR) which is also found in a number of other proteins that interact with various components of the complement system. These SCRs occur in the second component of complement, which is homologous to factor B, and also in the proteins C1r and C1s that are serine proteases involved in reactions at the beginning of the complement cascade.

The proteins in this large family have evolved to hydrolyse peptide bonds adjacent to certain amino acids only and consequently have many different specificities. This is because there are unique substrate binding sites in the enzymes that differ appreciably in structure between various members of the group.

One other interesting feature of this family is the great variability in the lengths of their introns. Some of the genes contain as little as 4% of coding information, while in others the proportion may be as high as 45%.

The histidyl–aspartyl–seryl triad is also used as the catalytic centre in some proteases from bacteria, such as subtilisin, and moulds. These enzymes must have evolved independently of the animal serine proteases since the critical active site residues are encoded in a different order – aspartyl, histidyl, seryl – and their crystal structures are quite different from those of trypsin and chymotrypsin. The genes for the bacterial enzymes, of course, have no introns, while the gene for the mould proteinase K from *Tritarchum* has only a single intron separating an exon encoding the aspartyl and histidyl residues from the one containing the seryl residue.

Some esterases, such as cholinesterase, also make use of a histidyl–aspartyl–seryl triad to effect their catalytic acts, but no genomic structures of these enzymes have yet been reported.

Figs. 11.2 and 11.3 make some general points about the structure of genes.

1. Genes coding for very similar proteins may contain very different numbers of exons.
2. Introns vary very widely in length, while the lengths of exons are more constant, rarely exceeding 300 nt.
3. Many genes have considerable numbers of untranslated nucleotides at the 3′-end. Rarely (not shown here), there may be introns interrupting these 3′-non-coding sequences.
4. Rather more commonly, there are introns interrupting the 5′-non-coding sequences. Though they are usually short, they are occasionally up to 1000 nt long (again not shown here).

11.4 Lipoproteins

The apo-lipoproteins are a group of proteins involved in the transport of water insoluble lipids in the aqueous medium of the blood. Seven (out of nine) are all closely related and must have arisen from a common ancestor. They all contain amphiphilic helices, with the hydrophobic side chains of amino acids projecting outwards on one face, presumably making contact with the lipids they transport, and hydrophilic amino acid side chains on the other face of the helix. Apo CII strongly activates lipoprotein lipase, which is responsible for the uptake of triacylglycerols from lipoproteins into tissues, while the closely related apo CIII inhibits the enzyme. Analogously, the enzyme lecithin-cholesterol acyl transferase, which has an important role in the transport of cholesterol, is activated by apo AI and inhibited by apo AII, although the two proteins are not so closely similar.

In all the genes, except that for apo AIV, there is a short non-coding exon at the 5′-end, and the three encoding apo AI, apo AIV and apo E all contain coding portions of their 3′-exons that are rather longer than usual (583, 1015 and 718 nt long). The genes for apo E, apo CI and apo CII are very close together on chromosome 19 in humans, and there is also a closely linked pseudogene for apo CI in which a C→T mutation has turned a glutamine codon (CAG) into a termination codon (TAG), so, if the protein were synthesised, it would almost certainly be non-functional. The genes for apo AI, apo AIV and apo CIII are also closely linked, this time on chromosome 11.

Apo B is a very large protein with no obvious resemblance to the other apoproteins. Its gene has 29 exons, one of which, with 7572 nt, is the longest known eukaryotic exon. The protein is synthesised in two forms which are generated from different mRNAs transcribed from the same gene. In the liver the full length of the gene codes for a protein of 4536 amino acid residues, while in the intestinal mucosa there is a post-transcriptional modification of a C residue to a U residue so that a codon for glutamine – CAA becomes UAA – a termination codon. This occurs at a position about half way along the mRNA so the protein has only 2152 amino acid residues.

11.5 Growth hormone family

The growth hormone family encompasses three hormones – Growth Hormone (GH) itself and prolactin, both secreted from the anterior pituitary, and somatomammotrophin (SMT) (also known as Placental Lactogen) which is secreted by the placenta, and may function as a growth hormone for the developing foetus. Prolactin and GH are believed to have arisen by gene duplication about 400 million years ago, while SMT probably diverged from GH much more recently. During the course of evolution the structure of human GH has diverged from that of non-primates to such an extent that the hormones from other mammals are inactive in humans, though the reverse is not true.

The rat has one gene for GH and one for prolactin. They both show a similar organisation of five exons and four introns, but all the introns in the prolactin gene are very much longer than those in the GH gene (Fig. 11.4). Thus, the prolactin gene is about five times as long as the GH gene, although they encode proteins of very similar size.

The human genome contains a single prolactin gene, again about five times the size of the GH gene, but there is a cluster of two GH and three SMT genes on chromosome 17. One of the GH genes encodes a protein that differs from normal GH at 15 positions in its amino acid sequence. This gene is not functional since it is present in the genome of an individual exhibiting GH deficiency, in whom the normal GH gene has been deleted.

Two forms of GH are found in human pituitaries, one of which has a deletion of amino acids 32–46. This sequence is encoded at the beginning of the third exon, and the variant form arises by an alternative form of splicing resulting in excision of the bases coding for these amino acids (Fig. 11.5).

Two of the SMT genes are expressed and encode proteins differing by only a single amino acid residue. There are four other base changes, but these are all silent in their coding properties. It is not known whether the other gene is expressed.

11.6 Glycoprotein hormones

The proteins of the glycoprotein hormone family are synthesised in the anterior pituitary. They regulate the functions of the thyroid gland

Fig. 11.4. The growth hormone (top) and the prolactin (bottom) genes of the rat.

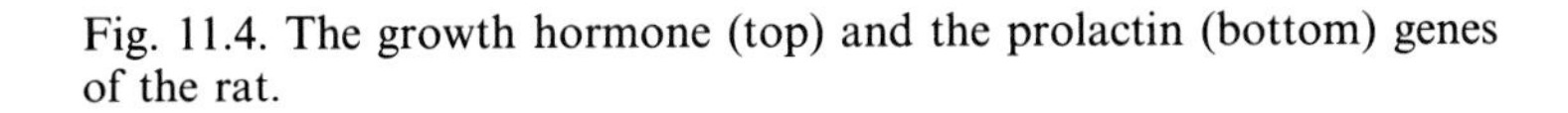

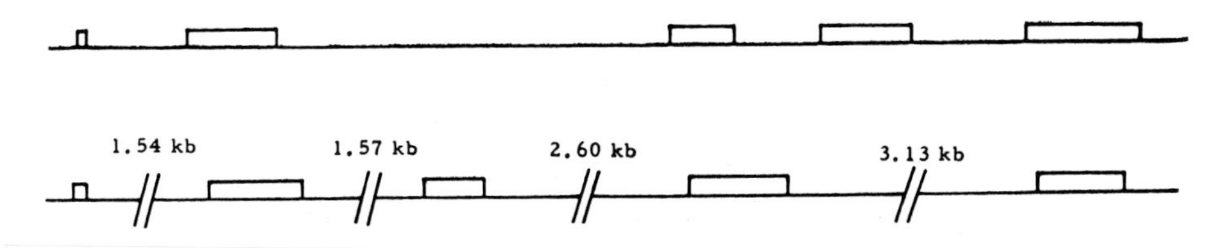

Table 11.1. *The glycoprotein family of hormones*

Hormone	Abbrev-iation	Site of production	Major sites of action
Thyroid-stimulating hormone, Thyrotrophin	TSH	Anterior pituitary	Thyroid
Follicle-stimulating hormone, Follitrophin	FSH	Anterior pituitary	Ovary and testis
Luteinising hormone, Lutrophin	LH	Anterior pituitary	Ovary and testis
Chorionic gonadotrophin	HCG	Placenta	Ovary

Note: The first three have alternative names, as shown.

and the gonads (Table 11.1). Another member of the family, chorionic gonadotrophin, is synthesised in the placenta of a few mammals and is homologous to LH in structure and function.

All these hormones consist of two non-covalently associated polypeptide chains. The α-subunit is common to all the hormones in a single species, while the β-subunits, though obviously similar, differ from each other and account for the biological specificity of the molecules. There is a small degree of homology between the mRNAs for the α- and β-subunits, suggesting that all the individual polypeptides have evolved from a common ancestor.

The β-subunit of HCG is 24 amino acids longer than the other β-subunits, and its mRNA has a very short 3′-flanking sequence in which the third, fourth and fifth nucleotides of the poly-adenylation signal (AAUAAA) are also used as the termination codon. Deletion of a single base pair a little way upstream from the termination codon of the β-subunit of LH during evolution has led to read through of this codon in

Fig. 11.5. Alternative splicing in human growth hormone mRNA. The upper line shows the sequence round the splice sites at the ends of the second intron which are used to produce the full-length hormone. The lower line shows the alternative 3′-splice site which is used to produce the shorter growth hormone molecules. Note that both 3′-splice sites are very similar, and show a high degree of homology with the consensus sequence (Table 7.11). The numbers are the numbers of the codons counted from the beginning of the mature hormone.

```
31                                               32
UUU GUAAGCUCU . . . . . . . . . UCCUUCUCCUAG GAA

31                                               47
UUU GUAAGCUCU . . . . . . . . . UCAUUCCUGCAG AAC
```

HCG (Fig. 11.6). Six β–HCG-like genes are present in the human genome, closely linked to each other and to that for LH, but only two are expressed.

11.7
Polyproteins

Some hormonally active polypeptides are synthesised as part of much larger proteins that are proteolytically processed to yield the active hormones. These precursors are known as polyproteins, despite the fact that the final products are generally peptides rather than proteins.

Pro-opiomelanocortin (POMC) is synthesised in the anterior pituitary. This is a precursor of several physiologically active peptides, including adrenocorticotrophin (ACTH), three melanotrophins (α-, β- and γ-MSH), and endorphin (Fig. 11.7). The gene encoding this polyprotein has one long intron in the 5′-flanking region and a second one in the protein-

Fig. 11.6. The structure of the 3′-terminal parts of the human genes for the β-subunits of LH (top) and HCG (bottom). The amino acids specified by the DNA sequences are shown above and below them. d = deletion; m = mutation.

```
   Gln Leu Ser Gly Leu Leu Phe Leu Ter
 --CAA CTC TCA GGC CTC CTC TTC CTC TAA---N71---AATAAA
     d   mm                      m
 --CG C TTC CAG GCC TCC TCT TCC TCA AAG---N69---CAA TAA A
   Arg Phe Gln Asp Ser Ser Ser Ser Lys          Gln Ter
```

Fig. 11.7. The structure of the primary translation product of the POMC gene (upper line). The short vertical lines show the positions of pairs of basic amino acid residues. The three lower lines show the products that can be formed by cleavage at some of these points. MSH = melanotrophin; ACTH = adrenocorticotrophin; LPH = lipotrophin; END = endorphin; CLIP = corticotrophin-like intermediate lobe peptide.

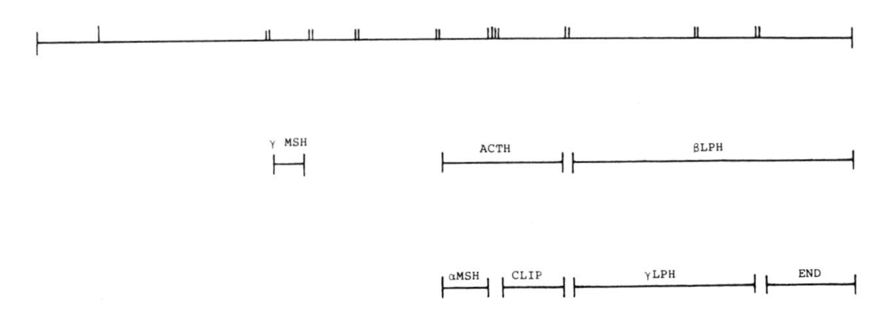

Table 11.2 *Ratio of products formed by alternative processing of POMC at different stages of development in the rhesus monkey pituitary*

Peptide	Foetus	Adult
ACTH	1.00	1.00
CLIP	1.21	0
α MSH	0.34	0
β MSH	1.01	0.17
β LPH	0.59	0.83
β END	1.71	0.98

Abbreviations as in Fig. 11.7.

coding sequence. Scattered throughout the protein are adjacent pairs of the two cationic amino acids – lysine and arginine – located at the beginning and end of the mature polypeptides (Fig. 11.7). This sequence of two basic amino acids acts as a signal for a peptidase found in tissues where precursor proteins are processed. The processing of POMC may occur in a number of ways in different parts of the pituitary or at different stages of development so that the peptides derived from it are formed in different proportions (Table 11.2).

The pentapeptide sequences of met-enkephalin and leu-enkephalin (Fig. 11.8) which are also found in some other peptides such as β-endorphin endows these compounds with potent analgesic properties. Some related peptides are synthesised from two larger precursors called prepro-enkephalins A and B.

The gene for preproenkephalin A from the adrenal medulla has introns in the 5′-flanking region and in the protein-coding sequence. The encoded protein contains no fewer than seven enkephalin sequences, some of which have C-terminal extensions. Preproenkephalin B is synthesised in the hypothalamus, and the protein it specifies contains three enkephalin peptides.

The two octapeptide hormones, vasopressin and oxytocin, are chemically very similar, but manifest very different physiological actions. They are synthesised in the hypothalamus and transported along axons in the pituitary stalk for storage in the posterior pituitary. During transport they are non-covalently bound to two homologous proteins – neuro-physins I and II. The synthesis of each octapeptide is directed by a gene

Fig. 11.8. The structure of leu-enkephalin (top) and met-enkephalin (bottom).

Tyr – Gly – Gly – Phe – Leu

Tyr – Gly – Gly – Phe – Met

coding additionally for the associated neurophysin. The octapeptides are split off the primary translation products at the usual signal dipeptide lysyl–arginyl sequence (Fig. 11.9). At the same time, the C-terminal residues of these peptides are amidated – a reaction involving a glycine residue that precedes the lysyl–arginyl pair (Fig. 11.10).

There are three exons in each gene, the first encoding a leader peptide, the hormone and part of the neurophysin. The second exon encodes most of the rest of the neurophysin, while the third exon in the oxytocin gene codes for the C-terminal end of neurophysin I. In the vasopressin gene, the third exon encodes both the C-terminal end of neurophysin II and a glycoprotein that functions as a prolactin-releasing hormone after it is cleaved from the neurophysin II post-translationally. Nearly all the second exon as well as 135 nt at the 3′-end of the first intron are identical in the two genes (Fig. 11.9). This complete conservation of a sequence of

Fig. 11.9. Bovine genes for the precursors of vasopressin and oxytocin and their protein products. Lines 2 and 3 show the structure of these genes, aligned to show the region of identity encompassing the 3′-end of the first intron and most of the second exon (indicated by vertical lines below the structures). Open rectangles are exons, with untranslated sequences lined vertically. Lines 1 and 4 show the structure of the protein products. The region of identity is again indicated by lines below the structures. The short vertical lines to the right of vasopressin and oxytocin indicate glycyl, lysyl and arginyl residues which are involved in the processing of the primary translation products. VP = vasopressin; OT = oyxtocin; NP = neurophysin; GP = glycoprotein.

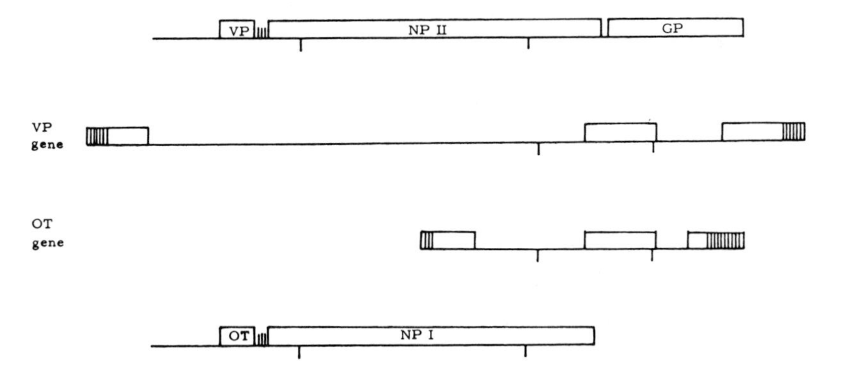

Fig. 11.10. Amidation of a peptide. The glycyl residue at the right-hand end of the precursor supplies the $-NH_2$ for the amidated product, and is converted to glyoxalate during the reaction which also requires molecular oxygen and ascorbate.

$$H_2N\text{-}R\text{-}CO\text{-}NH\text{-}CH_2\text{-}CO_2^- + O_2 \longrightarrow H_2N\text{-}R\text{-}CO\text{-}NH_2 + OHC\text{-}CO_2^- + H_2O$$

332 nt spanning an intron–exon junction is extremely unusual and is probably the result of a gene-conversion event that has occurred much more recently than the original divergence of the two genes.

The Brattleboro strain of rats lacks vasopressin and its associated neurophysin: consequently the animals suffer from diabetes insipidus so that they excrete very large volumes of dilute urine. The vasopressin gene in this strain has a single base pair deletion in the second exon (Fig. 3.16) that throws the rest of the reading frame out of phase, so there is no termination codon in the mRNA that is transcribed from the mutated gene. This must inhibit translation of the mRNA, since, while it is present in the hypothalamus of these rats, there are only minute amounts of its translation product.

12

Mitochondrial and chloroplast genomes

Mitochondria and chloroplasts contain single circular chromosomes that direct the synthesis of a number of proteins, and also the various RNAs that are required for this process.

12.1 Yeast mitochondrial genome

In yeasts, the mitochondrial genome is about 78 kbp long. Portions of it have been sequenced, but the complete sequence is not yet known. It encodes two rRNA molecules, a complete set of tRNAs, and mRNAs directing the synthesis of at least nine proteins (Fig. 12.1). The polymerases for the synthesis of mitochondrial DNA and RNA, all the tRNA synthetases and the majority of the functional mitochondrial proteins (e.g. enzymes of the citrate cycle and electron transport chain) are encoded by nuclear genes and synthesised on cytoplasmic ribosomes. All these proteins that are made in the cytosol contain amino acid sequences to target them for translocation into the mitochondria.

The mitochondrial rRNAs (21S (3200 nt) and 15S (1660 nt)) are somewhat smaller than prokaryotic rRNAs, and there is no rRNA corresponding to the 5·8S rRNA found in cytoplasmic ribosomes. The two rRNAs are encoded on widely separated portions of the genome, and are not transcribed together. There are genes for 24 tRNAs, some of which are clustered together, but others are found singly. There are separate tRNAs for N-formyl-methionine, which is used as an initiator amino acid as in prokaryotes, and for methionine when it is incorporated into the middle of a polypeptide. Otherwise, only threonine, serine and arginine have two tRNAs. The 5′-position in each anticodon is always either U or G, so that some codons are read by U–G pairing. When four codons are used for one amino acid, the 5′-base in the anticodon is always U. Presumably only the second and third bases of the anticodon are used when the third base in the codon is U or C (so-called 'two out of three' base pairing). The genetic code is slightly different from that which is usually used by nuclear genes, with UGA coding for tryptophan, rather than termination, while the codons beginning CU code for threonine (Table 12.1). There is a very strong bias against using codons with G or C in third position, in keeping with the unusual base composition of yeast mitochondrial DNA which has only 18% G+C.

The nuclear-encoded RNA polymerase is smaller and less complex than those used in nuclear transcription (Chapter 7.5), and is much more homologous to the RNA polymerases of certain bacteriophages than to that of *E. coli*. It consists of a single large core subunit that is directed to

Fig. 12.1. The yeast mitochondrial genome. For clarity it has been drawn twice. The inner circle shows the sites of the genes for the tRNAs, labelled according to the one-letter code for amino acids. The outer circle shows the exons as outwardly placed boxes and introns as inwardly placed boxes. 21S and 15S = rRNA; Cox I, II and III = subunits I, II and III of cytochrome oxidase; Cyt b = cytochrome b; ATPase 6 and 9 = subunits 6 and 9 of ATPase; RAP = ribosome-associated protein; U = open reading frames for which no corresponding proteins are yet known; m = maturase. There are also open reading frames in some of the introns. (After L. A. Grivell, 1983.)

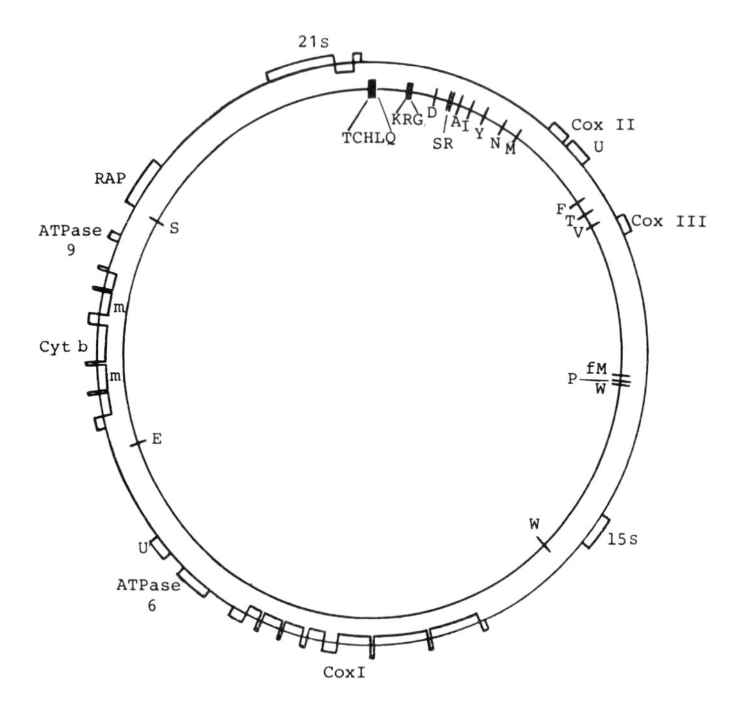

Table 12.1. *Differences in codon usage between mitochondrial and nuclear genes*

Codon	Amino acid encoded in: Nuclear genes	Yeast mitochondrial genes	Mammalian mitochondrial genes
CUU	Leu	Thr	Leu
CUC	Leu	Thr	Leu
CUA	Leu	Thr	Leu
CUG	Leu	Thr	Leu
UGA	Ter	Trp	Trp
AUU	Ile	Ile	Ile initiation?
AUA	Ile	Met	Met
AUG	Met	fMet	Met
AGA	Arg	Arg	Ter?
AGG	Arg	Arg	Ter?

Other codon usages are common to nuclear and mitochondrial genes.

Table 12.2 *Proteins encoded by yeast and mammalian mitochondrial genomes*

Subunits I, II and III of cytochrome oxidase
Apocytochrome b
Subunits 6, 8 and 9* of ATPase complex
var 1 (Ribosome associated protein)*
Maturases from introns 2 and 4 of apocytochrome b gene*
Endonuclease*
Subunits of NADH-coenzyme Q reductase†

* Only in yeast mitochondria
† Only in mammalian mitochondria

specific sites for initiation of transcription by a smaller protein specificity factor. It transcribes all the mitochondrial genes, irrespective of whether they encode rRNAs, tRNAs or mRNAs. The highly conserved sequence ATATAAGTA is used as a promoter for the initiation of all the transcripts that have been identified. The last A in this sequence is the actual base at which transcription starts. Many of the transcripts encode more than one gene product, so there must be secondary processing to produce the mature RNAs.

Several of the genes contain introns, but they are not invariably present. Some of the introns behave rather like transposable elements (Chapter 5.4). For example, crossing two strains of yeast, only one of which contains the intron in the 21S rRNA gene leads to the transfer of that intron into the intron-less gene. An endonuclease that is encoded in the intron recognises a specific sequence at the site of insertion and takes part in the reaction. A similar endonuclease is encoded in the fourth intron of the Cox I gene.

The genome encodes several proteins of the inner mitochondrial membrane (Table 12.2). The genes for the cytochrome oxidase subunit I and cytochrome b are both very much longer than needed to encode these proteins, and usually contain seven and five introns respectively. (Remember that introns are rare in genes of nuclear-encoded yeast proteins – Chapter 7.17.)

The cytochrome b gene occurs in two forms in different strains of yeast. In the short form the first four exons are already adjacent, with loss of introns I–III, while in other strains the long form has the full complement of six exons and five introns.

The processing of this gene has some interesting features. The excision of the first three introns is controlled by at least three nuclear-encoded genes. Removal of the first intron leaves an RNA molecule that is translated into a chimaeric protein. This is encoded by 143 codons from the first two exons of the gene plus another 280 codons from a continuous open reading frame that terminates near the 3′-end of intron II. This protein is known as a *maturase*, since it catalyses the excision of intron III – a stage in the maturation of the cytochrome b mRNA. It thus acts to destroy its own mRNA – a phenomenon known as splicing homeostasis. Another open reading frame in intron IV is translated into a second maturase that plays a part in the excision of its intron. The second maturase, as well as the products of the nuclear genes mentioned above, is involved in the excision of some of the introns of the Cox I gene. Some of these contain open reading frames that are more than 50 % homologous with the second maturase of the cytochrome b gene, and may also be involved in the processing of primary RNA transcripts.

The introns in these mitochondrial genes can be divided into two groups, each containing a unique set of highly conserved sequences that can assume common secondary structures (Fig. 12.2). Highly homologous structures are found in the corresponding genes in other fungal species, and also in nuclear genes coding for tRNA molecules in some *Protista*. None of these introns possess the highly conserved GU and AG sequences at their 5′- and 3′-ends that are found in nuclear-encoded introns (Chapter 7.17), and the mechanism of their removal by splicing is different from that which occurs in nuclear pre-mRNAs. These introns are spliced using the mitochondrially encoded maturases and the products of nuclear genes mentioned above.

Some of the Group II introns contain open reading frames with considerable homology to reverse transcriptase, though it is not certain that this activity is functional.

The small sizes of the ribosomal subunits and the small number of tRNAs required for protein synthesis in mitochondria suggest that there have been evolutionary constraints tending to keep the mitochondrial genome small, yet it is surprising that there appears to be a comparatively large proportion of this genome with no coding functions. It is also not clear why there has been selective pressure to maintain the separate synthesis of just a few mitochondrial proteins within the organelles. Those that are synthesised there are more hydrophobic than most known proteins: it is possible that, if these were made in the cytosol, their transport into the mitochondria might pose problems. However, this

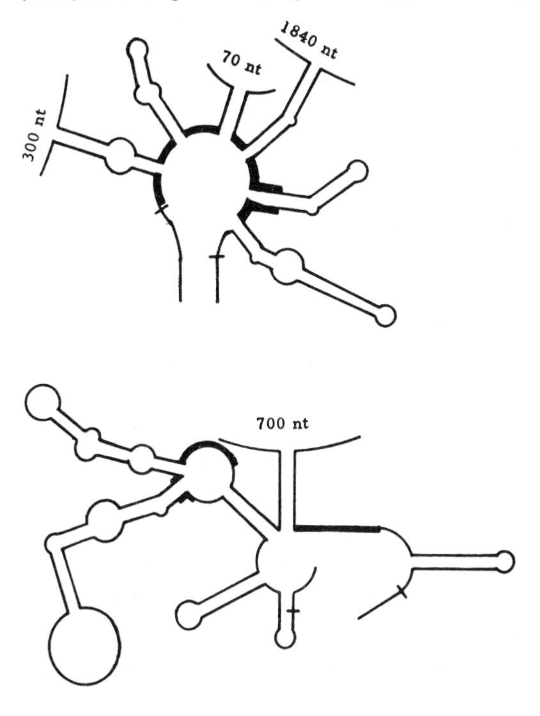

Fig. 12.2. Secondary structure of the two groups of introns found in fungal mitochondrial genes, and also in some *protistan* nuclear genes. The thickened lines represent very highly conserved sequences. The short lines across the main sequences denote the actual sites of splicing. The size of the large loops and some of the stems is somewhat variable. (Modified from F. Michel & B. Dujon. *EMBO J.* (1983), **2**, 33, figs. 2 & 3, reprinted by permission.)

cannot be the true or only explanation since, in *Neurospora crassa*, the most hydrophobic of these proteins, the ATPase proteolipid subunit 9, is synthesised in the cytosol.

Other fungi have smaller mitochondrial genomes – for example that of *Neurospora* is about 60 kbp long, and that of *Aspergillus* is only 33 kbp long. However, they probably encode more transcripts than the yeast genome. There are smaller intergenic sequences, and fewer (or no) introns in the genes homologous to those that do occur in yeast. The *Aspergillus* genome encodes some subunits of the NADH dehydrogenase complex that are not encoded in yeast mitochondria.

12.2 Mammalian mitochondrial genome

Human, bovine, mouse and rat mitochondrial genomes have been completely sequenced and they are all highly homologous and extremely similar in size (about 16.5 kbp). This small genome again only codes for a very limited set of proteins. Most of the mitochondrial proteins, including all those involved in mitochondrial transcription and translation, are encoded by the nuclear genome and synthesised on cytoplasmic ribosomes.

These mitochondria display an even more striking concentration of function into an extremely compact, highly utilised genome. The mitochondrial DNA can be separated into its two strands under denaturing conditions by taking advantage of their different densities due to their overall differences in base composition. These are referred to as the heavy (H) and light (L) strands. The map (Fig. 12.3) compiled from sequence data shows that most of the genetic information is encoded on the H strand: the L strand seems to code only for a few tRNAs and a single mRNA.

The mammalian mitochondrial genome has several features setting it apart from both nuclear and fungal mitochondrial genomes. It uses a slightly different genetic code, in which AUA, as well as AUG, codes for methionine and both are used as initiation codons; it does not use N-formyl-methionine for initiation. Other differences are shown in Table 12.1. Only 22 tRNAs are used, and when they have to read four different codons, U is always found at the 5′-end of the anti-codon. Leucine and serine are the only amino acids that require two tRNAs. There is a strong bias in favour of codons ending in A or C (77%), rather than in G or U.

This DNA is transcribed into one long precursor molecule corresponding to the full length of the genome. There are no normal flanking sequences for initiation of transcription of the mRNAs. They generally occur immediately adjacent to genes for some of the tRNAs which are believed to serve as punctuation signals. It seems plausible that the secondary structure that they adopt could direct an endonuclease to hydrolyse the newly transcribed RNA at appropriate places. A number of the mRNAs made in this way lack a complete termination codon at the 3′-end. However, they are all polyadenylated immediately after transcription, and this generates a suitable termination codon (UAA). They all lack the characteristic cap structure of most other eukaryotic mRNAs (Chapter 7.16).

Fig. 12.3. The human mitochondrial genome. For clarity is has been drawn twice. The two inner circles represent the H strand (outer) and the L strand (inner), and show the sites of the genes for tRNAs (labelled as in Fig. 12.1). The two outer circles show the sites of genes encoding proteins and rRNA. 12S and 16S = rRNA; cyt b = cytochrome b; Cox I, II, III = subunits of cytochrome oxidase; ATPase 6, 8 = subunits of ATPase; ND 1–6 and 4L = subunits of NADH dehydrogenase. A D-loop is shown at the top of the H strand. (After L. A. Grivell, 1983.)

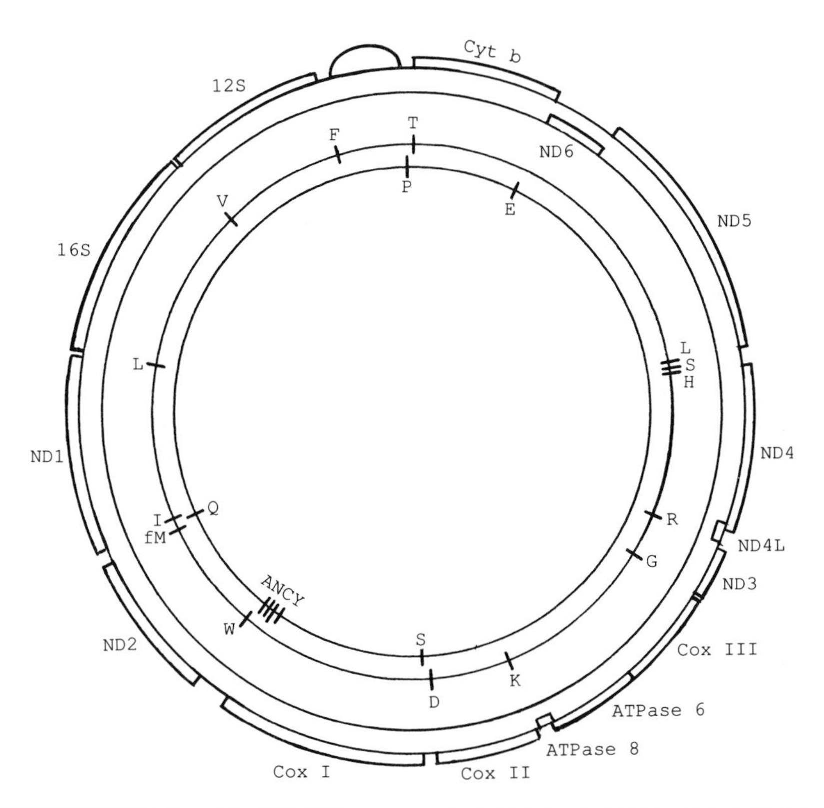

The H chain encodes 12 proteins containing between 68 and 603 amino acids. The genes for the two smallest (68 and 98 amino acids) overlap out of phase with those for two of the larger ones. A thirteenth polypeptide is encoded by the L chain. None of the mRNAs transcribed from these genes contains any introns. The proteins encoded are shown in Table 12.2. Over half the amino acids in these proteins have hydrophobic side chains, as would be expected since they are embedded in the inner mitochondrial membrane.

The mitochondrial rRNAs are much smaller than those found in the cytosol or in yeast mitochondria, with only 954 and 1559 nt (12S and 16S).

Most of the tRNAs have unusual structures with less base pairing than in the homologous cytosolic ones. They also vary in the length of the constant TψC loop (Fig. 2.5). Surprisingly, there is considerably less homology between the corresponding pairs of human and bovine tRNAs than between the corresponding pairs of cytosolic ones.

Only a small portion of the mammalian mitochondrial genome is not used for encoding RNAs, particularly the segment between the tRNAs for phenylalanine and proline. This is known as the *displacement* (*D*) *loop*, since a short length of DNA complementary to the L strand is synthesised, and this displaces the H strand so that the electron microscope reveals a D-shaped 'bubble' in the DNA. The proportion of molecules having this structure is somewhat variable and the function of the loop is unknown, though it may have a role in the control of the replication or transcription of the DNA. Synthesis of this short stretch of DNA requires an RNA primer that is initiated where transcription starts on the L strand. When the mitochondrial DNA is to be replicated, synthesis of the H strand starts here, but the start of L strand replication is far away, in the cluster of tRNA genes WANCY, and replication from this origin only starts when replication of the H strand is about 60% complete.

Transcription yields a large excess of rRNA over the individual mRNAs. This is achieved by the use of two different sites of initiation, one of which is used only for the transcription of the rRNA genes, while the other is used for transcribing the whole H strand.

The mitochondrial genomes of several other types of animals have been sequenced. That of *Xenopus* is extremely similar to mammalian ones, with all the genes in the same order, but it has a larger D loop. The mitochondrial genomes of sea urchins and *Drosophila* are slightly smaller, and though they encode the same number of mRNAs as their vertebrate counterparts these are arranged in a different order.

12.3 Plant mitochondrial genomes

The mitochondrial genomes of higher plants are very variable in length, even in members of the same family. For example, in the muskmelon the genome is about 2500 kbp long, while in the water melon its length is only about 200 kbp. The large sizes of these genomes has meant that, so far, none has been fully sequenced, but it seems that only a small portion actually codes for RNAs and proteins. Many plant genomes have directly repeated sequences 1–15 kbp long scattered throughout their length. These give rise to homologous recombination with the formation of smaller fragments of the genome, still in a circular configuration (Fig. 12.4). The significance of this is quite unknown.

Despite the large size of these genomes, they do not appear to encode many more RNAs than their fungal or animal counterparts. Three rRNAs of 5S, 18S and 26S are encoded, and in maize the 18S rRNA is actually slightly larger than the cytoplasmic nuclear encoded 18S rRNA. These rRNAs are more similar to bacterial and chloroplast rRNAs than to those of fungi or animals. A distinct set of tRNAs are also encoded and, probably, mRNAs for rather more proteins than are coded for by fungal

Fig. 12.4. Maize mitochondrial genome. The complete genome, 570 kbp in length is shown at the top. Some smaller fragments that are formed by homologous recombination of the repeated sequences (shown in black) are shown below. The figures refer to the length of the fragments in kbp. (After Dr. D. M. Lonsdale et al. (1984) Nucleic Acids Res. **12**, 9249, © 1984 IRC Press Ltd., by kind permission)

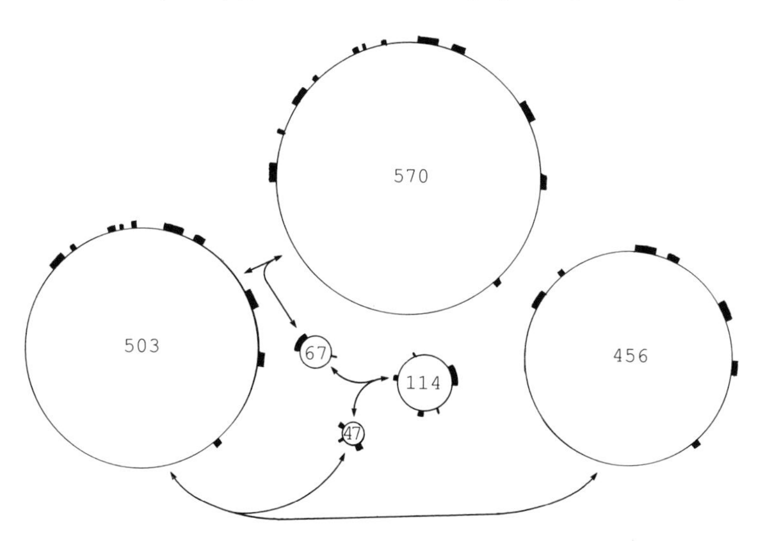

Table 12.3 *The sizes of some chloroplast genomes*

Species	kbp	Species	kbp
Geranium	217	*Chlamydomonas*	195
Nicotiana	156	*Pisum*	120
Marchantia	121	*Codium*	85

and animal mitochondrial DNAs. Those that have been identified include cytochrome b and subunits of ATPase and of cytochrome oxidase.

12.4 Chloroplast genomes

Chloroplasts, like mitochondria, contain a circular DNA genome that is generally about 100–200 kbp long (Table 12.3). The overall structure is remarkably well conserved over a very considerable evolutionary distance from a relatively simple liverwort (*Marchantia*) to several genera of higher plants. A genome of this size can encode over a hundred products but, like the mitochondrial genome, does not encode all the proteins that are used in the chloroplast.

The DNA typically contains two inverted repeats that encode rRNAs, some tRNAs and mRNAs (Fig. 12.5). The ribosomes are the same size as those of *E. coli*, and about a third of their proteins are encoded in the chloroplast genome. The 30 tRNAs synthesised in the chloroplasts are believed to read all the codons since a number can be read by 'two out of three' base pairing. There are genes for the three subunits of the RNA polymerase core enzyme, but no gene encoding a σ-like factor has been discovered. Nearly all the other identified proteins encoded in this genome are concerned with photosynthesis in one way or another.

Six of the primary transcripts for tRNAs and about ten of those encoding protein have introns, varying in length between 300–2500 bp.

Fig. 12.5. *Nicotiana* chloroplast genome. The thickened lines show the inverted repeats and the position and size of the rRNA genes are marked.

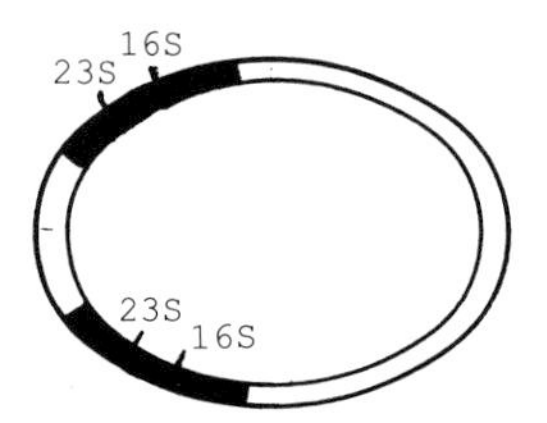

Some are similar to the Group I and II introns of fungal mitochondria, but some are of a third type. The mechanism(s) for their removal have not yet been clarified.

The sites of transcription initiation that have been identified are preceded by good homologies of the *E. coli* -10 and -35 sequences. Many mRNAs possess Shine–Delgarno-like sequences just 5′ to their initiation codons. In *in vitro* systems, *E. coli* RNA polymerase can transcribe a number of chloroplast genes. These facts all support the widely held theory that chloroplasts have been derived from endosymbiotic bacteria.

There are generally about 50 DNA molecules in each chloroplast so that an appreciable fraction of the cell's DNA may be present in these organelles. Little is known about the control of replication of these genomes, but the comparatively large amount of DNA present is likely to be required for the synthesis of new chloroplast constituents when cells divide.

13

Different and evolving genomes

13.1 The structure of prokaryotic and eukaryotic genes are different

Details of the structures and transcription of prokaryotic and eukaryotic genes can be found in Chapters 5 and 7, but the following summary highlights some of the major differences between the two groups of organisms.

The genes of both prokaryotes and eukaryotes have promoter sequences 5′ to the sites of initiation of transcription. Both contain a TATA box, though this generally starts about 10 nt 5′ to the transcription start site in prokaryotes and about 30 nt from this site in eukaryotes. In most yeast genes it is even further away, at 40–100 nt. It is not an invariant feature of eukaryotic promoters, but, as far as is known, it is always present in prokaryotes, where its actual sequence may influence the rate of initiation of transcription. TATA boxes that deviate from the consensus sequence are commonly found 5′ to genes that are only transcribed at a low rate so that there will be very low levels of the encoded proteins in each bacterium.

There is no sequence in eukaryotes corresponding to the prokaryotic −35 box, but eukaryotic genes typically possess a number of other upstream sequences to which trans-acting protein factors bind to regulate transcription.

The untranslated sequences 5′ to the initiation codon tend to be shorter in prokaryotes than in eukarotyes, where their length is very variable, ranging from just a few nucleotides to a thousand or more. In prokaryotes there is the conserved Shine–Delgarno sequence immediately 5′ to the initiation codon that is used to position the mRNA on the ribosome by base pairing. In eukaryotes, the methyl-guanosine cap at the extreme 5′-end of their mRNAs is required for binding to the ribosome. A well-conserved sequence immediately round the initiation codon presumably has a different role.

A very major difference between eukaryotic and prokaryotic genes is the almost universal presence of introns in the former. These could be relics of very early evolutionary time when primitive polypeptides may have been short. It is plausible that introns have been eliminated from prokaryotes by selection pressures leading to a more compact genome in these organisms. It is perhaps surprising that these genomes seem to have no relics of introns, though there are a few in the genomes of some bacteriophages. In this respect yeast is sometimes thought of as an 'honorary prokaryote' since introns are extremely rare, and their occurrence is limited mainly to genes encoding ribosomal proteins.

There are also differences at the 3′-end of genes. Most eukaryotic genes have the sequence AATAAA near their 3′-end where it acts as a polyadenylation signal. There is no such feature in prokaryotic genes where the mechanism for termination of transcription is probably very different from that used by eukaryotes.

13.2 Control mechanisms in prokaryotes and eukaryotes

There is considerable variation between the mechanisms of control of genetic expression in different types of organisms as well as a number of similarities.

In prokaryotes the rate of transcription is tightly controlled, generally by the concentration of metabolites. Since mRNA is unstable, its concentration is regulated mainly by the rate at which it is formed. Eukaryotic mRNA is usually much more stable so that its abundance changes more slowly than in prokaryotes, though its rates of formation can be varied widely in response to various stimuli. Even so, some mRNAs are degraded comparatively rapidly and there are specific sequences 3′ to the termination codon that label the mRNAs for this purpose. Rapid changes in metabolism are brought about post-translationally, for example, by allosteric effects on enzymes or by their covalent modification. These mechanisms are also widely used in prokaryotes.

A very major difference between the two kingdoms is in the organisation of many bacterial genes into operons, while there is no corresponding arrangement in eukaryotes. However, as was noted in Chapter 6, the genes encoding enzymes that all catalyse reactions on a particular metabolic pathway (e.g. arginine biosynthesis) are sometimes scattered throughout the bacterial chromosome but are still under unified control. This is exerted through sequences to the 5′ side of the genes.

The transcription of some bacterial genes may be controlled through the binding of cAMP to the Catabolite Activating Protein which, in turn, binds to sites on the DNA. Some eukaryotic genes are also responsive to the protein CREB when it has bound cAMP. It is perhaps not surprising that the DNA sequences to which these proteins bind show little similarity.

Eukaryotic promoters frequently contain a number of binding sites for different regulatory proteins. These may be involved in responses to tissue specific elements or hormones, either by direct binding of a hormone, or by binding a second messenger produced by the primary action of the hormone. Metabolites may also bind to these regulatory proteins in unicellular eukaryotes such as yeast, but it is doubtful whether this form of control is used in multicellular organisms. Zinc fingers and leucine zippers (Chapter 7.9) are probably used widely as control elements in eukaryotes only, whereas the helix-turn-helix motif figures prominently in control in both prokaryotes and eukaryotes.

More than 20 prokaryotic DNA binding proteins with the helix-turn-helix motif have been characterised, but it is also found in an interesting set of eukaryotic proteins that have very important roles in cell differentiation and the development of the organism. It has been known for over 70 years that *Drosophila* contains genes that control its segmental development. These genes are known as homeotic genes and the proteins that they encode all possess a highly conserved element of about 60 amino acids called the *homeobox*. There are very similar genes in other eukaryotes – *Xenopus*, mammals, and, more distantly related, in yeasts. The genes are clustered in complexes, suggesting that they evolved by duplication and mutation. The homeobox proteins contain in addition to the helix-turn-helix motif, a longer and reasonably well-conserved helix on the N-terminal side of this. The C-terminus tends to have several amino acids with anionic side chains, giving it an overall negative charge. There are also two other small regions of homology towards and at the N-terminus.

The conservation of particular residues is stronger between the eukaryotic proteins than in the prokaryotic ones. Perhaps the latter regulate a wider range of functions. The eukaryotic homeotic proteins can be grouped into families on the basis of conserved amino acid sequences.

In *Drosophila* the transcription of the various protein members of this family is localised to specific regions of the developing embryo. Similar localisation is also found in early vertebrate embryos, though the distribution of these proteins becomes more widespread as the embryo develops.

Regulatory proteins generally possess distinct domains, each of which has a specific function, such as the DNA binding site, the ligand binding

Table 13.1. *The properties of chimaeric proteins that have been made by swapping DNA binding sites and activating regions of regulatory proteins of different species*

DNA binding site	Activating region	Result
LexA*	GCN4†	Activates transcription in yeast cells when bound to either LexA or GCN4 binding sites
LexA	c-fos‡	Activates in yeast cells when LexA binding site is engineered 5′ to yeast genes
GAL4†	Human estrogen receptor	Activates in HeLa cells§ in hormone dependent manner
Human estrogen receptor	GALI†	Activates transcription in yeast cells in hormone dependent manner

* Bacterial protein.
† Yeast protein.
‡ Human protein.
§ Cell line derived from human tissue.

site and activating or repressor sites that, in some cases, are believed to interact with RNA polymerase or accessory transcription factors to influence transcription directly. It is possible to engineer chimaeric proteins possessing domains from different sources and demonstrate that they are still active. For example, the hormone binding domain of the zinc finger proteins that bind glucocorticoids and thyroxine can be swapped with predictable effects on hormone action on the genes that they normally control. More extreme examples of this are the swapping of domains between bacterial, yeast and vertebrate regulatory proteins. Some examples are shown in Table 13.1 and in Fig. 13.1.

These experiments involve several DNA binding proteins, including the yeast regulatory proteins GAL4 and GCN4, vertebrate steroid hormone receptors and the bacterial protein LexA. This is produced in *E. coli* and normally represses the transcription of an enzyme involved in the repair of damaged DNA. Lex A has a helix-turn-helix motif near its N terminus that binds to a specific DNA sequence.

In these experiments the chimaeric protein is generally expressed from a suitably constructed plasmid that is transfected into the cell under investigation, together with a second plasmid containing a reporter gene (Chapter 3.10) that is placed under the control of an appropriate DNA sequence that binds the chimaeric protein.

The results of these experiments suggest that when these proteins are bound to their respective binding sites on DNA their activating regions are brought into close proximity with some elements of the transcription apparatus – either RNA polymerase II, or some of the accessory transcription factors. However, the actual nature of the binding site is not at all critical, provided that the activating protein can bind to it, and it is in a suitable location relative to the transcriptional start site. The binding sites have generally been positioned 5′ to the promoters of the reporter genes, but in one case the chimaeric protein was bound in an intron rather than 5′ to the transcription start site. The activating domains of these proteins are able to act across species boundaries since a vertebrate steroid hormone receptor can activate a yeast gene in a hormone-dependent manner when the receptor binds to its DNA element placed 5′ to a yeast promoter.

Fig. 13.1. Chimaeric constructs of LexA DNA binding site and rat glucocorticoid receptor. The shaded portions are the DNA binding sites of the proteins. H is the hormone binding site of the glucocorticoid receptor. The results shown were obtained in a rat cell line when co-transfected with a plasmid containing the LexA binding site inserted 5′ to a reporter gene.

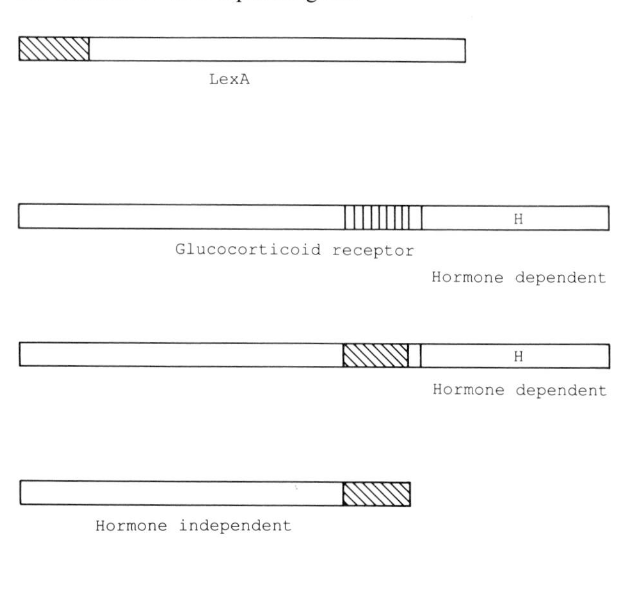

13.3 Repeated sequences occur widely in many genomes

It is estimated that only about 1–2% of the human genome contains coding sequences that are ever expressed. Most of the rest is made up of various kinds of repeated sequences. The *Drosophila* genome also has a high content of repetitive sequences whose lengths are frequently around 5 kbp, and a number of repeated sequences have been discovered in the genomes of higher plants. In contrast, the fungi seem to have more compact genomes with only 5–10% of moderately repetitive DNA.

There are two major kinds of these repeated sequences in eukaryotes that can frequently move about the genome. Transposons are bounded by inverted terminal repeats and move by excision and re-integration. The other kind, known as *retroposons*, integrate into the genome after reverse transcription of the RNA that the elements encode. This latter kind exhibits a close analogy with bacterial transposons on the one hand (Chapter 5.4), and with retroviruses on the other (Chapter 8.1).

About half of the repeated sequences in the human genome are known as SINES (short interspersed sequences), with an average length of 400 bp, organised into families each encompassing about 10^5 members. In addition, there are families of LINES (long interspersed sequences) that are usually up to 6 kbp long. In the few cases where they have been analysed, individual families contain about 10^4–10^5 members.

However, the most abundant of the repeated sequences in the human genome are known as the Alu family because they all possess a site cleaved by the restriction endonuclease Alu I. A few individual ones have been sequenced and their sequences, though not identical, are highly conserved.

Fig. 13.2. Alu sequence in the human (top) and rodent (bottom) genomes. D = Direct repeat at insertion site of the sequence; A = runs of dA nucleotides; H = human specific insert; R = rodent specific insert.

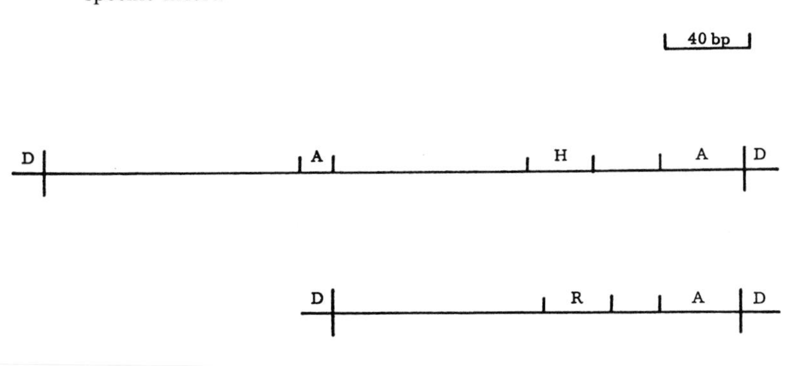

They are about 300 bp long, and there are 300000–500000 copies in each cell comprising about 3% of the human genome, and occurring, on average, at intervals of about 5000 bp. They are composed of two repeated sequences of 130 bp, one of which has an insert of 31 bp. There are similar structures in rodent genomes, though these only possess a single copy that corresponds to the repeating unit in the human sequences plus a non-homologous 32 bp insert (Fig. 13.2). Both human and rodent Alu sequences have dA runs of up to about 40 nucleotides at the 3′-end, and both families are flanked by repeated sequences up to 19 bp long. There is some evidence that there are related families in other vertebrates.

Since many of the longer introns contain Alu sequences they are major constituents of hnRNA (Chapter 2.3.3). These sequences contain promoters that are recognised by RNA polymerase III, and so can be transcribed from any position in the genome since this enzyme uses sequences in the DNA it transcribes as promoters. In fact, transcripts of Alu sequences have been found.

Human 7SL RNA (Chapter 2.3.4) has a region that is about 80% homologous to the Alu family. This RNA is used in the removal of the signal peptide that is present on the precursors to all secreted proteins (Chapter 7.19). A 4·5S RNA, found in rodents, but not in humans, also shows considerable homology to rodent Alu sequences, but it has no known function.

It is believed that these Alu sequences may be moved about the genome in which they occur, but, apart from encoding the 7SL and 4·5S RNAs, they do not seem to have any functions. It has been suggested that they might be regarded as 'parasitic' DNA that does not contribute to the phenotype of the organism.

In *Drosophila* there are at least 20 families of repeated sequences, each occurring about 30 times. The positions at which they occur vary in different strains of fly, and even between individuals of the same strain. To some extent they are randomly inserted throughout the genome, though a definite preference is shown for certain sites. Their widespread distribution has led to the suggestion that they are highly mobile elements that can sometimes take adjacent DNA sequences with them when they are moved around the genome. There has been genetic evidence for this behaviour for many years, and it is gratifying that this can be integrated with evidence obtained from studies at the molecular level.

In yeasts, the Ty 1 family that has similar characteristics has been studied extensively, and there are a number of mobile elements in the genomes of higher plants.

These repeated sequences are flanked by direct terminal repeats that are usually about 250–500 bp long, with much shorter inverted repeats at their extreme ends that can form the stem of a stem and loop structure by base pairing. There is a short repetition (3–5 bp) of the genomic base sequence on each side of the sites where these transposons are integrated into the genome, presumably because a staggered cut is made in the target DNA prior to insertion.

There are about 50 copies of the P element in *Drosophila* in male germ line cells of some strains. It is nearly 3 kbp in length and contains two open reading frames. RNA transcripts have been detected so that it could direct the synthesis of two proteins. Smaller versions of the P element also occur in which internal deletions have removed parts or all of the open reading frames; in fact they are generally more abundant than the intact P elements.

When a male fly whose genome contains P elements is crossed with a female fly that does not, the P elements that are transmitted to the offspring move around the genome and are inserted more or less at random. This process is catalysed by a transposase that is believed to be encoded by the element itself. Once this has happened, the P elements are usually stable, probably because the protein produced from the other open reading frame acts as a transcriptional repressor.

The P element has been isolated and used as a vector to transfer genes into the DNA of *Drosophila*. Several thousand base pairs can be spliced into a P element which has been inserted into a plasmid that is then transfected into flies where some of the modified P elements will integrate into the host's chromosomes and express the inserted DNA.

13.4 The plasticity of the genome

One of the most striking findings that have emerged from recent studies of the genome is that it is continually in a dynamic state and not a fixed stable body of information that is always handed on intact from one cell to its descendants. This is largely because there are a number of ways in which recombination between the various portions of genomic DNA can take place. Insertion sequences and transposons move around the genome, often taking neighbouring stretches of DNA with themselves, and they can be inserted at sites where they may disrupt and inactivate various functional DNA sequences. Perhaps more occasionally they can induce the transcription of certain DNA sequences in circumstances in which they are not normally transcribed. Gene conversions and crossovers can generate new genes and link together genes that were previously unlinked. Translocations between chromosomes can cause drastic effects in activating genes that are not normally transcribed, and are probably one of the mechanisms by which cancers arise. Finally, viruses can be integrated into the host chromosomes and subvert the metabolism of the infected cell through the production of virally encoded products. Many of these changes, particularly those occurring in somatic cells, may never be observed since they may be so deleterious to the cell in which they occur that these affected cells do not survive, or they may reproduce so slowly that the effects are swamped out by more rapidly growing 'normal' cells. Effects on the genome of gametes are more likely to be observed, though even here, where many million spermatozoa die for each one that actually fertilises an ovum, the chances of observing changes are extremely low.

Occasionally, advantage has been taken of this plasticity of the genome as in the production of the vast number of antibodies of differing specificity that can arise by random reshuffling of the cistrons that go to make up the complete H and L chain genes (Chapter 10.2).

When the protozoal trypanosomes parasitise a host, there is rapid and repeated recombinatorial scrambling of a family of cistrons encoding their surface antigens. As fast as the host generates antibodies to react with a particular antigen, a new and different set of antigenic determinants is synthesised. A considerable time must elapse before complementary antibodies to these can be produced which will inactivate the parasite.

In prokaryotes the plasticity of the genome is rather more evident since they usually divide more rapidly than eukaryotic cells so that favourable mutations can spread rapidly through a population. It is comparatively easy, for example, to select bacteria to grow on an unusual nutrient by cultivating them continuously in the presence of steadily increasing concentrations of it. Equally, resistance to toxic substances such as antibiotics is fairly rapidly acquired, especially through the transmission of plasmids, and this has become a major hazard especially in hospitals throughout the more developed countries.

So it seems that the genome is subject to considerable changes and modifications and it is probably valuable that there are considerable portions (such as introns and non-coding sequences) which provide a pool of material of potential value for the production of new materials which can be selected by natural selection if they are found to have survival value.

13.5 Evolution

The plasticity of the genome provides an explanation for the evolution of species, regardless of the means by which this is driven. Comparison of the sequences of homologous proteins (or, better still, the genes encoding them) from different species can suggest the pathways by which present-day proteins have evolved. Where there is good palaeontological evidence of the time of divergence of species it is possible to compare this with the rate of accumulation of mutations in the DNA to provide a molecular clock for evolution. Introns and non-coding portions of the genome which are not involved in the controlling functions provide the best parts for the study of mutation rates. A recent estimate of this rate is about 1 mutation in 2×10^8 sites per year. Even in the DNA actually coding for proteins mutations may be silent because of the redundancy of the genetic code. In certain portions of most proteins constraints on the precise structure may not be very great. Mutations which conserve the properties of the amino acids that are encoded (e.g. leucine for isoleucine or valine; glutamate for aspartate) are commonly found. However, in certain critical regions intimately concerned with the protein's function very little change can be tolerated. These regions tend to be highly conserved, and will evolve at a much slower rate (if at all) than other regions.

In situations where the structure of a large number of homologous

Fig. 13.3. Tree showing the probable evolutionary path of some κ and λ chains of immunoglobulins. The circles represent points at which lineages diverge. The linear distances between them are proportional to the number of changes in amino acids that have occurred. (Modified from L. Hood. *Fed. Proc.* (1976), **35**, 2162, Fig. 6, reprinted with permission.)

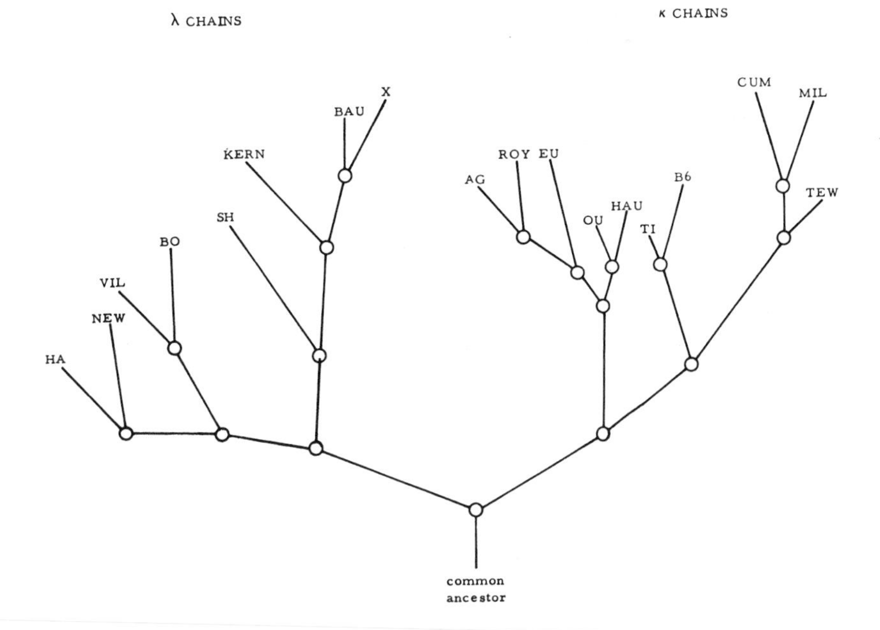

Table 13.2. *Rates of evolutionary change in the amino acid sequence of some proteins*

Protein	(*a*)
Histones	
H4	400
H3	330
Structural proteins	
Collagen	36
Crystallin	22
Intracellular enzymes	
Glutamate dehydrogenase	55
Lactate dehydrogenase	19
Cytochrome c	15
Carbonic anhydrase	4
Extracellular enzymes	
Trypsin	6
Ribonuclease	2.3
Hormones	
Glucagon	43
ACTH	24
Insulin	14
Insulin C peptide	1.9
Growth hormone	4
Serum proteins	
Albumin	3
Immunoglobulin V regions	0.7–1.0
Immunoglobulin C regions	0.9–1.7
Snake venoms	0.8–0.9

(*a*) The numbers refer to the unit evolutionary period, the numbers of millions of years for 1% change in sequence.

proteins or genes is available it is possible to construct trees showing the routes by which they have probably evolved. A specially favourable case is that of the antibody family of proteins (Fig. 13.3).

The overall rates of evolution observed in different proteins are strikingly different (Table 13.2). At one extreme are the histones H3 and H4, while at the other are the immunoglobulins and snake venoms. There are probably two major factors accounting for these wide differences. Conservation of function is obviously important and presumably limits the changes that can be tolerated in proteins like enzymes with highly specific functions. In this context, it is interesting that there has been wide diversification of some classes of enzymes which have identical catalytic centres with very strong conservation of amino acids directly involved in the catalytic act. Neighbouring parts of the molecules which function as binding sites for the substrates have evolved much faster and led to the ability to catalyse reactions of different substrates. This is seen, for example, in the large family of serine proteases (e.g. trypsin, chymotrypsin, elastase), and also in the phosphokinases where the binding site for ATP has been conserved while that for the second substrate varies quite widely from enzyme to enzyme.

Dispensability is also an important factor. Serum albumin is an excellent example of this. It has evolved relatively rapidly, and to a large extent, its functions can be taken over by other plasma proteins, since in the admittedly rare condition of analbuminaemia the total lack of the protein does not seem to lead to impaired viability. Snake venoms and immunoglobulins are also free to evolve rapidly provided that toxicity and antigen-binding properties respectively are maintained. In fact, diversification of the latter is likely to be advantageous in conferring protection against a wider range of noxious molecules.

Gene duplications have been a common feature during evolution. This is especially evident in duplications of a number of globin genes (Chapter 9.1). Similar events must have given rise to the various forms of the H chains of antibodies. Here it is of interest to note how the H chain clusters in the mouse and human genomes have undergone duplications in different ways (Fig. 10.5).

Recent theories of the origins of life have postulated that the first oligomeric molecules could well have been similar to RNAs. The sugar moieties might have been trioses or their derivatives which are simpler than ribose. The self-splicing of the 26S rRNA intron of *Tetrahymena* and other introns supports the idea that molecules other than proteins can have catalytic activity, like the RNA component of RNase P. It is quite possible that other reactions such as removal of nuclear encoded introns and peptidyl transferase could be catalysed by RNAs rather than by proteins. Several enzymes are ribonucleoproteins and in some cases their activity is sensitive to digestion with RNase, suggesting the importance of RNA for the catalyses. The earliest proteins to evolve may have oriented substrates in the most favourable configuration for catalysis by primitive RNAs. Proteins, with their more varied amino acid side chains than the four bases in RNA probably provide more versatile structures and greater potential for catalytic acts than RNAs, so the evolution of catalytic proteins would obviously have been advantageous.

13.6 Future developments

With the advent of rapid and reliable DNA sequencing it is relatively easy to determine the sequence of large DNAs. At present about 20% (1000 kbp) of the *E. coli* chromosome has been sequenced, and sequences are being added at the rate of about 200 kbp per year. It is likely that it will not be too long before the complete sequence of this genome is known.

There are also projects under way to sequence all the yeast chromosomes and there is the much more ambitious Human Genome Project which, it is hoped, will provide the sequence of the whole of the human genome by the end of the century. According to present estimates there are about 10^5 genes in this genome which would probably require about 10^8 bp of coding sequence. Since the human genome contains a total of $5{\cdot}6 \times 10^9$ bp it is likely that only about 2% or less is used for coding purposes. This suggests that there are long stretches with no apparent function. Powerful computer routines are being developed to store all the sequences and to detect those that may be of particular interest.

Knowledge of the sequences of all protein coding genes will provide useful tools for the diagnosis of heritable diseases. Eventually, it is possible that ways may be found to introduce fully functional genes into individuals who have inherited defective ones.

Many unexpected and fascinating concepts and mechanisms have come to light as a result of experimental work on the genome, and it is almost certain that Nature still has a number of surprises in store for us as we gain further information.

Glossary

Blunts ends The termini produced when double-stranded DNA is cleaved by an enzyme which hydrolyses the strands between nucleotides that are precisely opposite each other.

Cap The methyl-guanylate residue that is added post-translationally to most eukaryotic mRNAs.

cDNA (Complementary DNA) DNA that is synthesised by reverse transcriptase using an RNA template.

Chromosome walking The joining up of restriction maps (qv) by the use of overlaps.

Codon A triplet of three consecutive bases in DNA specifying an amino acid.

Complementation The restoration of a function that has been lost by a functional copy of the appropriate gene.

Cross over Reciprocal exchange of DNA between two chromosomes.

Domain Part of a protein which is responsible for a particular property of the molecule.

Downstream Used to denote sequences 3′ to a given sequence.

Enhancer DNA sequence to which protein(s) can bind to enhance the rate of transcription of a nearby gene. Distinct from the promoter (qv).

Eukaryotes Organisms containing cells with separate nuclei.

Exon A sequence of DNA within a gene whose complement appears in mRNA.

Gene conversion Unidirectional transfer of DNA from one chromosome to a similar site elsewhere.

Genetic engineering The production (and usually expression) of DNA derived from different genomic sites.

Genome The complete DNA of one organism.

Heat shock proteins Proteins that are produced in increased amounts when cells are exposed to various insults such as growth at high temperatures.

Heterogeneous nuclear RNA (hnRNA) Newly transcribed RNA in the nucleus prior to processing.

Homeobox A sequence of about 60 amino acid residues, well conserved across species barriers, which is part of a larger protein involved in embryonic differentiation.

Intron A sequence of DNA that is transcribed into RNA but later removed.

Mutation A change in nucleotide sequence in DNA.

Oncogene A mutated gene that induces unrestricted proliferation when taken into a normal cell.

Open reading frame A DNA sequence that can be translated into a polypeptide sequence with no termination codons except at the C-terminus.

Operator A region of DNA, adjacent to the promoter, which participates in the control of the activity of the latter.

Passenger DNA DNA that is engineered into a vector so that it can be amplified.

Phenotype An observable heritable characteristic of an organism.

Point mutation A change in a single nucleotide in a DNA sequence.

Prokaryote An organism whose cells have no separate nucleus.

Promoter A region of DNA, 5′ to a gene, that is required for correct initiation of transcription.

Pseudogene A mutated form of a gene, still in the genome, that is either not transcribed, or, if it is, produces a non-functional protein.

Replication The production of new DNA.

Replication fork The region of double-stranded DNA where it is opened up during replication.

Restriction map A map of DNA showing the position of sites where restriction enzymes can act.

Signal peptide A sequence of amino acids at the N-terminus that directs a newly synthesised polypeptide into the endoplasmic reticulum, where it is cleaved off prior to secretion of the mature protein.

Splicing The process in which introns are eliminated from a pre-mRNA transcript.

Sticky ends Short, unpaired termini of DNA produced when double-stranded DNA is cleaved by an enzyme that hydrolyses the two strands between nucleotides that are not opposite each other.

Transcription The synthesis of RNA on a DNA template.

Transfection The introduction of DNA into vectors or cells.

Translation The synthesis of a polypeptide directed by mRNA.

Upstream Used to denote sequences 5′ to a given sequence.

Vector DNA that is capable of replicating itself when introduced into a suitable host cell into which passenger DNA (qv) can be inserted.

Reading lists

Chapter 1

Bauer, W. R., Crick, F. H. C. & White, J. H. (1980) Supercoiled DNA. *Sci. Amer.* **243(1)**, 100

Cozzarelli, N. R. (1980). DNA gyrases and the supercoiling of DNA. *Science*, **207**, 953

Travers, A. A. (1989). DNA conformation and protein binding. *Ann. Rev. Biochem.* **58**, 427

Travers, A. A. & Klug, A. (1987). DNA wrapping and writhing. *Nature, London*, **327**, 280

Wang, J. C. (1985). DNA Topoisomerases. *Ann. Rev. Biochem.* **54**, 665

Wells, R. D. (1988). Unusual DNA structures. *J. Biol. Chem.* **263**, 1095

Zimmerman, S. B. (1982). The three-dimensional structure of DNA. *Ann. Rev. Biochem.* **51**, 395

Chapter 2

Yonath, A. & Wittman, H. G. (1989). Challenging the three-dimensional structure of ribosomes. *Trends Biochem. Sci.* **14**, 329

Kim, S-H. (1978). Three dimensional structure of transfer RNA and its functional implications. *Adv. Enzymol.* **46**, 279

Schimmel, P. R., Soll, D. & Abelson, J. (eds). (1980). *Transfer RNA*. Cold Spring Harbor, NY: Cold Spring Harbor Laboratory

Birnstiel, M. L. (ed.) (1988). *Structure and Function of Major and Minor Small Nuclear Ribonucleoprotein Particles*. Springer-Verlag

Kozak, M. (1987). An analysis of 5′-noncoding sequences from 699 vertebrate messenger RNAs. *Nucleic Acid Res.* **15**, 8125

Lake, J. A. (1985). Evolving ribosome structure. *Ann. Rev. Biochem.* **54**, 507

Moore, P. B. (1988). The ribosome returns. *Nature, London* **331**, 223

Chapter 3

Watson, J. D., Tooze, J. & Kurtz, D. T. (1983). *Recombinant DNA – A Short Course*. Scientific American Books, New York: W. H. Freeman & Co.

White, T. J., Arnheim, N. & Evlin, H. A. (1989). The polymerase chain reaction. *Trends Genet.* **4**, 185

Jordan, B. E. (1988). Megabase methods: a quantum jump in recombinant DNA techniques. *Bioessays* **8**, 140

Gilbert, W. (1981). DNA base sequencing and gene structure. *Bioscience Rep.* **1**, 353

Sanger, F. (1981). Determination of nucleotide sequences in DNA. *Bioscience Rep.* **1**, 3

There are several articles on Nucleases in *The Enzymes, vol. XIV, A.* (1981). (ed. P. D. Boyer). New York: Academic Press

Chapter 4

Bolivar, F. & Backman, K. (1979). Plasmids of *E. coli* as cloning vectors. *Methods Enzymol.* (ed. R. Wu), **68**, 245

Collins, J. (1979). *Escherichia coli* plasmids packageable *in vitro* in bacteriophage particles. *Methods Enzymol.* (ed. R. Wu), **68**, 309

Blattner *et al.* (1977). Charon phages: safer derivatives of bacteriophage lambda for DNA cloning. *Science*, **196**, 161

Hohn, T. & Katsura, I. (1987). Structure and assembly of bacteriophage lambda. *Curr. Topics Microbiol. Immunol.* **78**, 69

Burke, D. T., Carle, G. F. & Olson, M. V. (1987). Cloning of large segments of exogenous DNA into yeast by means of artificial chromosome vectors. *Science*, **236**, 806

Anderson, W. F. & Diacumakos, E. G. (1981). Genetic engineering in mammalian cells. *Sci. Amer.* **245(1)**, 60

Miller, L. K. (1989). Insect baculoviruses: powerful gene expression vectors. *Bioessays.* **11**, 91

A number of articles in *Methods Enzymol.* (1987), **153**, (ed. R. Wu & L. Grossman) discuss recent advances in the use of vectors.

Chapter 5

Dale, J. W. (1989). *Molecular Genetics of Bacteria.* Chichester: Wiley

Glass, R. E. (1982). *Gene Function.* E. coli *and its Heritable Elements.* London: Croom Helm

Kornberg, A. (1980). *DNA replication.* San Francisco: W. H. Freeman & Co. and its supplement. (1982). San Francisco: W. H. Freeman & Co.

McHenry, C. S. (1988). DNA polymerase III of *E. coli. Ann. Rev. Biochem.* **57**, 519

Bear, D. G. & Peabody, D. S. (1988). The *E. coli* Rho protein: an ATPase that terminates transcription. *Trends Biochem. Sci.* **13**, 343

Helmann, J. D. & Chamberlin, M. J. (1988). Structure and function of bacterial σ factors. *Ann. Rev. Biochem.* **57**, 839

McClure, W. R. (1985). Mechanism and control of transcriptional initiation in prokaryotes. *Ann. Rev. Biochem.* **54**, 171

Platt, T. (1986). Transcription termination and the regulation of gene expression. *Ann. Rev. Biochem.* **55**, 339

Travers, A. A. (1987). Structure and function of *E. coli* promoter DNA. *Crit. Rev. Biochem.* **22**, 181

Papers in *Cold Spring Harbor Symposium on Quantitative Biology.* (1980), **45**

Chapter 6

Miller, J. H. & Reznikoff, W. S. (eds.) (1980). *The Operon.* Cold Spring Harbor, NY: Cold Spring Harbor Laboratory

Yanofsky, C. (1981). Attenuation in the control of bacterial operons. *Nature, London*, **289**, 751

Lindahl, L. & Zengel, J. M. (1986). Ribosomal genes in E. coli. *Ann. Rev. Genet.* **20**, 297

Nomura, M. (1984). The control of ribosome synthesis. *Sci. Amer.* **250(1)**, 72
Gallant, J. A. (1979). Stringent control in *Eschericia coli. Ann. Rev Genet.* **13**, 393

Chapter 7

Isenberg, I. (1979). Histones. *Ann Rev. Biochem.* **48**, 159
Kedes, L. H. (1979). Histone genes and histone messengers. *Ann. Rev. Biochem.* **48**, 837
Wu, R. S. *et al.* (1987). Histones and their modifications. *Crit. Rev. Biochem.* **20**, 201
Kornberg, R. D. & Klug, A. (1981). The nucleosome. *Sci. Amer.* **244(2)**, 48
Morse, R. H. & Simpson, R. T. (1988). DNA in the nucleosome. *Cell*, **54**, 285
Campbell, J. L. (1988). Eukaryotic DNA replication. *Trends Biochem. Sci.* **13**, 733
Kelly, T. J. (1988). SV40 DNA replication. *J. Biol. Chem.* **263**, 17889
Stillman, B. (1988). Initiation of eukaryotic DNA replication *in vitro*. *Bioessays* **9**, 56
Atchison, M. L. (1988). Enhancers; mechanism of action and cell specificity. *Ann. Rev. Cell Biol.* **4**, 127
Bird, A. P. (1987). CpG islands as gene markers in the vertebrate nucleus. *Trends Genet.* **3**, 342
Dynan, W. S. (1989). Modularity in promoters and enhancers. *Cell*, **58**, 1
Elgin, S. C. R. (1988). The formation and function of DNase hypersensitive sites in the process of gene activation. *J. Biol. Chem.* **263**, 19259
Geiduschek, E. P. & Tocchini-Valentini, G. P. (1988). Transcription by RNA polymerase III. *Ann. Rev. Biochem.* **57**, 873
Green, S. & Chambon, P. (1988). Nuclear receptors enhance our understanding of transcription. *Trends. Genet.* **4**, 309
Hinnebusch, A. G. (1988). Mechanism of gene regulation in the general control of amino acid biosynthesis in *Saccharomyces cerevisiae*. *Microbiol. Rev.* **52**, 248
Johnson, P. F. & McKnight, S. L. (1989). Eukaryotic transcriptional regulatory proteins. *Ann Rev. Biochem.* **58**, 799
Johnston, M. (1987). A model fungal gene regulatory mechanism: the GAL genes of *Saccharomyces cerevisiae*. *Microbiol. Rev.* **51**, 458
Lehman, I. R. & Kaguni, L. S. (1989). DNA polymerase α. *J. Biol. Chem.* **264**, 4265
Maniatis, T., Goodbourn, S. & Fischer, J. A. (1987). Regulation of inducible and tissue-specific gene expression. *Science*, **236**, 1237
Mitchell, P. J. & Tjian, R. (1989). Regulation in mammalian cells by sequence-specific DNA binding proteins. *Science*, **245**, 371
Mowry, K. L. & Steitz, J. A. (1988). snRNP mediation of 3′-end processing. *Trends Biochem. Sci.* **13**, 447
Proudfoot, N. J. (1989). How RNA polymerase II terminates transcription in higher eukaryotes. *Trends Biochem. Sci.* **14**, 105
Ptashne, M. (1988). How eukaryotic transcriptional activators work. *Nature, London*, **335**, 683.
Sentenac, A. (1985). Eukaryotic RNA polymerases *Crit. Rev. Biochem.* **18**, 31
Sollner-Webb, B. & Tower, J. (1986). Transcription of cloned eukaryotic ribosomal genes. *Ann. Rev. Biochem.* **55**, 861
Struhl, K. (1989). Molecular mechanisms of transcriptional regulation in yeast. *Ann. Rev. Biochem.* **58**, 1051
Breitbart, R. E., Andreadis, A. & Nadal-Grinard, B. (1987). Alternative splicing. *Ann. Rev. Biochem.* **56**, 467
Padgett *et al.* (1986). Splicing of messenger RNA precursors. *Ann. Rev. Biochem.* **55**, 1119

Chapter 8

Barbacid, M. (1986). Mutagens, oncogenes and cancer. *Trends Genet.* **2**, 188

Barbacid, M. (1987). *ras* genes. *Ann. Rev. Biochem.* **56**, 779

Bishop, J. M. (1983). Cellular oncogenes and retroviruses. *Ann. Rev. Biochem.* **52**, 68

Land, H., Parada, L. F. & Weinberg, R. A. (1983). Cellular oncogenes and multistep carcinogenesis. *Science*, **222**, 771

Leder, P. *et al.* (1983). Translocations among antibody genes in human cancer. *Science*, **222**, 765

Varmus, H. E. (1982). Form and function in retroviral proviruses. *Science*, **216**, 812

Chapter 9

Efstratiadis, A. *et al.* (1980). The structure and evolution of human globin genes. *Cell*, **21**, 653

Flavell, R. A. (1983). The globin genes of rabbit and man. *Biochem. Soc. Trans.* **11**, 111

Weatherall, D. J. & Clegg, J. B. (1981). *The Thalassaemia Syndromes*. 3rd edn Oxford: Blackwell Scientific Publications

Vella, F. (1980). Human haemoglobins and molecular disease. *Biochem. Educ.* **8**, 41

Chapter 10

Kay, J. et al (eds.) (1986). Genes and proteins in immunity. *Biochem. Soc. Symp. No. 51*

Honjo, T. (1983). Immunoglobulin genes. *Ann. Rev. Immunol.* **1**, 499

Tonegawa, S. (1983). Somatic generation of antibody diversity. *Nature, London* **302**, 575

Williams, A. F. & Barclay, A. N. (1988). The immunoglobulin superfamily – domains for cell surface recognition. *Ann. Rev. Immunol.* **6**, 381

Clevers, H. *et al.* (1988). The T cell receptor/CD3 complex: a dynamic protein assembly. *Ann. Rev. Immunol.* **6**, 629

Davis, M. M. & Bjorkman, P. J. (1988). T-cell antigen receptor genes and T-cell recognition. *Nature, London*, **334**, 395

Flavell, R. A. *et al.* (1986). Molecular biology of the H-2 histocompatibility complex. *Science*, **233**, 437

Kapper, M. & Strominger, J. L. (1988). Human Class II MHC genes and proteins. *Ann. Rev Biochem.* **57**, 991

Chapter 11

Hirt, H. J. *et al.* (1987). The human growth hormone gene locus: structure, evolution and allelic variations. *DNA* **6**, 59

Lauer, S. J. *et al.* (1988). Two copies of the human apolipoprotein C-1 gene are linked closely to the apoliprotein E gene. *J. Biol. Chem.* **263**, 7277

Nebert, D. W. & Gonzalez, F. J. (1987). P-450 genes: structure, evolution and regulation. *Ann. Rev. Biochem.* **56**, 945

Talmadge, K., Vamkapopoulus, N. C. & Fiddes, J. C. (1984). Evolution of the genes for the subunits of human chorionic gonadotrophin and luteinising hormone. *Nature, London*, **307**, 37

Richter, D. (1983). Vasopressin and oxytocin are expressed as polyproteins. *Trends Biochem. Sci.* **8**, 278

Chapter 12

Anderson *et al.* (1981). Sequence and organisation of the human mitochondrial genome. *Nature, London*, **290**, 475
Attardi, G. (1985). Animal mitochondrial DNA. *Internat. Rev. Cytol.* **93**, 92
Grivell, L. A. (1983). Mitochondrial DNA. *Sci. Amer.* **248(3)**, 60
Michel, F. & Dujon, B. (1983). Conservation of RNA secondary structure in two intron families including mitochondrial-, chloroplast- and nuclear-encoded members *EMBO J.* **2**, 33
Umesono, K. & Ozeki, H. (1987). Chloroplast gene organisation in plants. *Trends Genet.* **3**, 281
Newton, K. J. (1988). Plant mitochondrial genomes. *Ann. Rev. Plant Physiol.* **39**, 503

Chapter 13

Jelinek, W. R. & Schmid, C. W. (1982). Repetitive sequences in eukaryotic DNA and their expression. *Ann. Rev. Biochem.* **51**, 813
O'Hare, K. & Rubin, G. M. (1983). Structure of P transposable elements and their sites of insertion and excision in the *Drosophila* genome. *Cell*, **34**, 25
Schmid, C. W. & Jelinek, W. R. (1982). The Alu family of dispersed repetitive sequences. *Science*, **216**, 1065
Spradling, A. C. & Rubin, G. M. (1982). Transposition of cloned P transposable elements into *Drosophila* germ line chromosomes. *Science*, **218**, 341
Papers in *Cold Spring Harbor Symposium on Quantitative Biology*. (1980), **45**, 519–673
Cech, T. R. & Bass, B. L. (1986). Biological catalysis by RNA. *Ann. Rev. Biochem.* **55**, 599
Joyce, G. F. (1989). RNA evolution and the origin of life. *Nature, London*, **338**, 217
Wilson, A. C., Ochman, H. & Proger, E. M. (1987). Molecular time scale for evolution. *Trends Genet.* **3**, 241

Index

F refers to figures; T refers to Tables